Impacts of Diesel-Powered Light-Duty Vehicles

Diesel Cars
Benefits, Risks, and Public Policy

Final Report of the
Diesel Impacts Study Committee

Assembly of Engineering
National Research Council

NATIONAL ACADEMY PRESS
Washington, D.C.   1982

   This report represents the work performed under contract No.
68-01-5972 with the U.S. Environmental Protection Agency, the U.S.
Department of Energy and the U.S. Department of Transportation.

Library of Congress Catalog Card Number 81-86631
International Standard Book Number 0-309-03237-7

Available from:

NATIONAL ACADEMY PRESS
2101 Constitution Ave., N.W.
Washington, D.C. 20418

Printed in the United States of America

NATIONAL RESEARCH COUNCIL

OFFICE OF THE CHAIRMAN
2101 CONSTITUTION AVENUE
WASHINGTON, D.C. 20418

December 21, 1981

The Honorable Ann M. Gorsuch
Administrator
Environmental Protection Agency
Washington, D.C.   20460

Dear Mrs. Gorsuch:

With this letter I transmit to you the report entitled Diesel Cars: Benefits, Risks, and Public Policy, prepared by a committee of the National Research Council, under Contract 68-01-5972 with the U.S. Environmental Protection Agency, the U.S. Department of Energy, and the U.S. Department of Transportation.

This report is the latest in a series on the subject of motor vehicle emissions that the National Research Council has issued in the past seven years. The reports were all based on studies undertaken pursuant to Section 403 (f) of Public Law 95-95 in which the Congress has directed the Administrator of the Environmental Protection Agency "to enter into appropriate arrangements with the National Academy of Sciences to conduct continuing comprehensive studies and investigations of the effects on public health and welfare of emissions subject to Section 202 (a) of the Clean Air Act ... and the technological feasibility of meeting emission standards required to be prescribed by the Administrator."

The study that concludes with this report elucidates the scientific, technological, and economic bases for making regulatory policy for diesel cars. One of the principal conclusions of the report is that diesel passenger cars and lightweight trucks, in their current numbers at least, do not appear to present a threat to health and the environment-- though the report emphasizes that our knowledge about diesel emissions is not definitive and proposes various approaches of research to strengthen our knowledge.

The committee responsible for this report suggests that the
air quality and fuel economy goals of the nation, which your
agencies are concerned with, should not be compromised by reducing
or eliminating the current regulatory standards for diesel cars,
and that more rigorous standards are not likely to be necessary
for the 1985 model year. It further urges the EPA to closely
monitor the results of present and future health and environmental
research and, in 1983, and perhaps every three years afterward,
assess the validity of diesel emission standards--particularly
in light of the numbers and locations of diesel cars at the time.

In addition to these recommendations, the committee calls
upon the Congress and the EPA to consider regulating heavy diesel
trucks and buses, which emit greater amounts of particulate matter
than light-duty diesel vehicles. This, the committee observes,
is likely to be a more cost-effective strategy than more stringent
regulation of diesel cars and light trucks.

I want to express the appreciation of the National Research
Council to Henry Rowen and the members of the committee and panels
for their contributions to this difficult and sensitive study.

Yours sincerely,

Frank Press
Chairman

Identical letters sent to:

The Honorable James B. Edwards
Secretary of Energy

The Honorable Drew Lewis
Secretary of Transportation

The Honorable George A. Keyworth II
Director, Office of Science and Technology Policy

NATIONAL RESEARCH COUNCIL
ASSEMBLY OF ENGINEERING

2101 Constitution Avenue   Washington, D. C. 20418

DIESEL IMPACTS STUDY COMMITTEE

202/389-6811
202/389-6974

December 7, 1981

Dr. Frank Press, Chairman
National Research Council
2101 Constitution Avenue, N.W.
Washington, D.C.   20418

Dear Dr. Press:

It is my pleasure to submit to you the complete final report, Diesel Cars: Benefits, Risks, and Public Policy, prepared by the Diesel Impacts Study Committee.

When the Committee began its study in the summer of 1979, the nation was in the grip of yet another shortage of motor fuel, brought on by the turbulent revolution in Iran, with concomitant increases in oil prices. Once again, as in the 1973-1974 period, when we were cut off from petroleum produced in several Middle East countries, the conservation of oil was a central theme in government and throughout society. As a way of saving fluid fuel, some Americans were turning to diesel-powered passenger cars. At the same time that diesel cars were gaining acceptance, questions were being raised about the possibility that diesel engine emissions could damage human health and reduce air quality. Though the US Environmental Protection Agency already had imposed certain limits on the various pollutants emitted by diesel vehicles, the prospective growth in the number of diesel cars and the questions this presented led the White House Office of Science and Technology Policy to ask the National Research Council to assess the situation.

The study was undertaken to assist the executive and legislative branches of the government, as well as the American people in general, to better understand the benefits and costs of the wider use of diesel cars and small trucks. It was clear from the start that the Committee's purpose was to examine the scientific, technical, and economic factors and to provide an analytic base that could be of considerable value to those who must formulate government policy about diesels. Beyond this, the Committee's report would help the public to perceive the benefits and risks of "dieselization" of America's roads.

Our first concern, and that of the sponsoring agencies, was to assess the evidence on the potential health hazards of diesel emissions. This was done in the form of an Interim Report by the Health Panel of the Committee last year.

On the broader question of the social issues involved in "dieselization," we found that very little was known about the overall costs and benefits of government regulation to protect health, safety, and environmental amenities. Our job involved putting societal and individual advantages and disadvantages into perspective, comparing the hazards and the alternatives to determine where and on whom the greatest risks would fall. In this way we could bring more light to the decision-making process for regulating diesel engine emissions. After all, there is no scientific formula for making regulatory decisions. There is no satisfactory way to calculate all the costs and benefits of regulatory alternatives in dollars or any other terms that can be mathematically added, subtracted, or compared.

There is no single, objective, definitive policy that the Committee agreed was tenable for all time so long as the answers to the questions about diesel emissions and their control remain imprecise. Notwithstanding all the uncertainties, the Committee has provided some significant assistance to regulators and legislators. The report offers some findings, however tentative, about whether diesel emissions are any more critical to health than known carcinogens in cigarette smoke, say, or roofing tar. It emphasizes that present knowledge is far from complete and points to ways that the knowledge base can be strengthened by additional research. It suggests when the regulatory agency ought to make checks of the research in order to apply the knowledge to setting appropriate standards for disesl emissions.

Because of the diverse backgrounds and viewpoints represented on the Committee, as well as the complexity of the subject, effective communication and consensus building were not always easy. And because the members are not equally expert in all aspects of the problem, there may be no member of the Committee who agrees with every detail of this report. But all the members agree with all of the essential conclusions of the report.

As chairman, I want to express my appreciation to the Committee members for their contributions to the study and to the panels and consultants who prepared so many documents and summaries to advance this endeavor. Finally, the Committee owes a great debt to Irwin Goodwin of the staff of the National Research Council for holding us together in the final months, seeking consensus, and editing and producing this report. He brought to our work the requisite combination of enthusiasm, patience, cheerfulness, and, most of all, the important quality of caring.

Sincerely,

Henry S. Rowen

Henry S. Rowen

DIESEL IMPACTS STUDY COMMITTEE

Henry S. Rowen, <u>Chairman</u>
Professor of Public Management
Stanford University

William M. Capron
Professor of Economics
Boston University

Kenneth S. Crump
President
Science Research Systems, Incorporated

Alan Q. Eschenroeder
Senior Staff Scientist
Arthur D. Little, Incorporated

Sheldon K. Friedlander
Vice-Chairman
Department of Chemical, Nuclear, and Thermal Engineering
University of California at Los Angeles

Bernard M. Goldschmidt
Associate Professor of Environmental Medicine
New York University Medical Center

Herschel E. Griffin*
Associate Director and Professor of Epidemiology
Graduate School of Public Health
San Diego State University

Jack D. Hackney
Professor of Medicine
Rancho Los Amigos Hospital (University of Southern California)

*Former Dean of the Graduate School of Public Health
 University of Pittsburgh

STAFF

L. F. Barry Barrington
    Executive Director

Claudia Anderson
    Program Coordinator

S. Ann Ansary
    Program Coordinator

Scott R. Baker
    Staff Officer

Dennis Miller
    Senior Staff Officer

Claude E. Owre
    Editor

Kenneth R. Smikle
    Staff Officer

Rudolph C. Yaksick
    Staff Officer

Vivian Scott
    Administrative Secretary

Juliet W. Shiflet
    Secretary

Julia W. Torrence
    Secretary

Michele Zinn
    Secretary

This report presents the results of a comprehensive study of the risks
and benefits associated with the wider use of diesel-powered motor cars
and light trucks and the implications for regulating such vehicles.
The National Research Council undertook the study in May 1979 at the
request of the White House Office of Science and Technology Policy,
with the support of the U.S. Environmental Protection Agency, the U.S.
Department of Energy, and the U.S. Department of Transportation.

The report is the latest in a series of scientific and
technological assessments by committees of the National Research
Council on the subject of motor vehicle emissions. The first of these
was issued in January 1972. That report and others published through
December 1974 were based on studies directed by the U.S. Congress in
the Clean Air Act of 1970 and its later revisions. The earlier studies
were concerned principally with the technological feasibility of
meeting the statutory standards established for light-duty motor
vehicle emissions.

The present report deals with dimensions of the problem of motor
vehicle emissions that were not as forcefully evident in the early
1970's--notably, public demand for diesel-powered passenger automobiles
and small trucks with high fuel efficiency to offset the rising price
of motor fuels and government concern that diesel exhaust fumes contain
particulates and chemicals that may be harmful to humans and their
environment. Indeed, as diesel engines replace conventional gasoline
engines in new automobiles and lightweight trucks, the United States is
experiencing one of the most important changes in the history of
automotive technology.

The study was undertaken to inform the three government bodies,
along with the U.S. Congress, the automobile industry, and the American
public, about diesel engine emissions and their control as well as the
other implications of a large increase in the number of light-duty
diesel vehicles. All things considered, the government organizations
responsible for regulating and administering for health, safety, and
energy need the scientific, technological, and economic data and
analyses on which to base their policy decisions.

Accordingly, the National Research Council organized the Diesel
Impacts Study Committee in the Assembly of Engineering, which operated
in conjunction with the Assembly of Life Sciences for aspects of the

study dealing with the possible health hazards of diesel engine emissions. The scope of the study is defined in the contract between the Environmental Protection Agency and the National Academy of Sciences as (1) an examination of the current state of experimental and theoretical research on adverse health effects from diesel emissions, including an evaluation of the limitations and implications of such work, taking into account the technological feasibility of reducing emissions, and (2) an analysis of the comparative risks and benefits of expanding the use of diesel-powered automobiles and light trucks on the nation's roads.

The committee consisted of 20 members drawn from diverse disciplines and backgrounds--medical research, health care, environmental protection, chemical and mechanical engineering, political science, economics, banking, and business management. Because the study involved a complex range of issues and interactions, the committee established four panels to examine, respectively, the aspects involving technology, environment, human health, and public policy. Each of the panels was made up of specialists drawn from the relevant area of concern as well as some members of the committee. In performing their separate tasks the panels sometimes called on experts to assist in examining special matters and explicating particular problems. The members of each panel are listed in Appendix A, along with consultants and contributors to the study.

The committee's first meeting took place in July 1979 at the National Academy of Sciences' Summer Study Center in Woods Hole, Massachusetts. A second summer workshop was held at Woods Hole in August 1980 to review a draft of the final report. In between, the panels visited manufacturing companies and research centers in the United States and abroad. In December 1979, the Health Effects Panel, with substantial contributions from the Analytic Panel and the full committee, prepared a report for the Environmental Protection Agency in the form of a letter reviewing an assessment of the potential carcinogenic impact of diesel engine exhaust. Two panels also prepared separate reports to inform the committee. Valuable documents in their own right, one of the reports, Health Effects of Exposure to Diesel Exhaust, was issued in October 1980, and the other, Diesel Technology, is forthcoming. Both reports, like the committee's own report, were reviewed under procedures approved by the Report Review Committee of the National Research Council. In addition, the committee has caused three commissioned reports to be published as supporting papers. These are cited in the present report and listed in Appendix B.

The collection of data and the deliberation of findings and conclusions were completed for the most part at the time of the committee's meeting in August 1980, although the committee met in October and December 1980 to consider drafts of this report. In May 1981, the committee chairman and the panel chairmen met to respond to the comments about this report by representatives of the Report Review Committee and the Assembly of Engineering. While some statistical data have been gathered to make the report more timely, it needs to be emphasized that the conclusions and analyses were developed during 1980. The committee acknowledges that the health research and environmental studies reported here have continued and that some new work has been undertaken. The committee also recognizes that the technologies of diesel engines and emission controls have advanced in the past year. Notwithstanding those developments, the committee considers the conclusions in this report to be valid and useful.

During the course of the study, three individuals resigned from different panels. Jose A. Gomez-Ibanez, of the Department of City and Regional Planning at Harvard University, left the Analytic Panel to join the staff of the President's Council of Economic Advisers; Jane V. Hall, of the California Air Resources Board, left the Technology Panel when she joined Union Oil Corporation; and James N. Pitts, Jr., Director of the Statewide Air Pollution Research Center, University of California at Riverside, resigned from the Health Effects Panel after disagreeing with the way the summary and conclusions of the panel's report were presented. The committee is grateful to these people for their incisive ideas expressed in the first year of the study.

SUMMARY AND CONCLUSIONS

Invented in Germany in 1892, the diesel engine has been a sturdy, reliable, and efficient power system for decades for ships and submarines, railroad locomotives, buses, heavy trucks, farm tractors, and earth-moving equipment. By contrast, the diesel in passenger cars and small trucks has been in low gear. Now, at least one U.S. car maker projects that by 1990 diesels will account for one in every four new automobiles and lightweight trucks it produces. The reason for the acceleration in diesel sales is its fuel mileage savings (from 25 to 35 percent greater than current gasoline-powered vehicles of equal size and weight). This is of obvious benefit to car owners in an era when petroleum is no longer cheap and may not always be abundant, as well as to car makers in meeting the federal requirement that their total output of 1985 model light-duty vehicles must average 27.5 mpg or better.

Like all motor vehicles burning petroleum products, diesels are a source of air pollution. While federal regulations limiting some tailpipe emissions have been in effect for more than a decade, standards for diesels have been promulgated only recently, mainly because of the prospective increase of such vehicles on the nation's roads. Diesel emissions differ from gasoline engine exhaust in several ways: Diesels meet the current emission standards for hydrocarbons and carbon monoxide (CO), set under the Clean Air Act and its amendments (42 USC 7401 et seq.), but produce somewhat more nitrogen oxides (NO_x), sulfur dioxide (SO_2), and aldehydes; they also emit from 30 to 100 times more particulate matter by mass than gasoline engines fitted with catalytic converters. Particulate emission rates for light diesels range from 0.2 g/mi (grams per mile) for the smallest cars to 0.7 or 0.8 g/mi for the largest, while gasoline-powered cars and small trucks with catalytic converters discharge only about 0.02 g/mi from their tailpipes.

Diesel particulates are of special concern. Bits of aggregated carbon formed during incomplete combustion of diesel fuel, they are submicron in size, with complex organic chemicals adsorbed on their surface. Among the many chemicals in diesel exhaust are known or suspected carcinogens, toxic substances, and mutagens. The soot particles are so small that they can elude the body's normal respiratory defenses and deposit deep within the lungs where they may lodge for weeks or longer. What is more, if particulates in the atmosphere are present in abundance, visibility can be reduced considerably.

Other emissions also produce effects. NO_x decreases human resistance to disease and affects the atmospheric ozone balance in a way that is not completely understood. SO_2 as well as NO_x contribute to a phenomenon known as "acid rain." Aldehydes irritate human lungs and induce the formation of photochemical smog.

To reduce the possible risks of health hazards and air pollution from diesel cars and light trucks, the U.S. Environmental Protection Agency established standards in 1979 for the control of particulate emissions. As orginally set forth, the standard for the 1981 and 1982 models years was 0.6 g/mi, and this would be tightened to 0.2 g/mi for 1983 and thereafter. The agency set the dates in the belief that adequate time existed for all manufacturers to develop emission control technology, though it recognized that some diesel cars weighing about 2,000 pounds and operating with small displacement engines would probably meet the 0.2 g/mi particulate level before 1983 without emission control devices, by improvements and modifications to engines. When some car manufacturers asserted they could not meet the lower level of particulates by 1983, however, the EPA postponed the adoption of the 0.2 g/mi standard to 1985.

Concerned about the implications of diesel cars and emission regulations for the health of the general public and the automotive industry, the EPA, along with the U.S. Departments of Transportation and Energy, called on the National Research Council in mid-1979 to perform a detailed examination of the available information and understanding about diesel emissions, to review the technological aspects of controlling diesel exhaust, and to provide a risk-benefit assessment of the government's regulatory standards. Accordingly, the National Research Council's Diesel Impacts Study Committee has reached the following major findings and conclusions.

CONTROL TECHNOLOGY

A late starter for passenger cars and small trucks, the diesel engine is less developed today than the gasoline engine and, thus, its potential is great for technological advances to further its fuel economy, improve its performance, and reduce its tailpipe emissions. The advances are most likely to come in fuel injection systems, combustion chamber design, electronic engine controls and sensors, turbochargers, and exhaust gas recirculation (EGR).

Even so, the development of diesel engines with low particulate emissions will be constrained by engine fuel quality and control systems, as well as by NO_x standards. EGR, the primary NO_x emission control technique in current use for diesels, is known to increase the quantity of particulates. Currently, then, there is a trade-off between NO_x and particulate emissions. Thus, meeting the light-duty vehicle emission standard for particulates in the 1982-1984 model years is technically feasible if some waivers of the NO_x standard are granted in 1983 and 1984 for the largest cars, pickup trucks, and vans.

Complying with the NO_x standard of 1.0 g/mi and the particulate standard of 0.2 g/mi for light-duty diesels in 1985 is technologically feasible for only those vehicles weighing less than 2,000 pounds. Limiting the particulate level to 0.2 g/mi for diesels of more than 2,000 pounds most likely will require an exhaust after-treatment device such as a trap oxidizer. None of the particulate control devices now under development has yet been proven in road durability tests of 50,000 miles. The history of developing the catalytic converter for gasoline engines in the early 1970's suggests about a five-year lag between the design and demonstration of such devices and their production and commercialization--an interval that will make it difficult for the larger and heavier diesel cars and pickup trucks to meet the EPA's 1985 standard.

The composition of diesel fuel also affects particulate emissions, but producing fuel of higher quality, which could reduce pollutants by a small amount, may limit the availability of diesel fuel. Today's refinery capacity for diesel fuel is sufficient to satisfy the needs of the most likely number of diesel passenger cars and small trucks that may be introduced through at least 1990. Still, the quality of diesel fuel now varies considerably. By setting minimum specifications for such fuel for cars, trucks, and buses, engineers would be able to improve engine designs and ensure that the performance of existing diesels does not degrade.

Because large diesel-powered trucks and buses, individually and collectively, emit much greater amounts of particulate matter than small diesels, EPA should consider controlling the particulate levels of heavy-duty diesels. In addition, the agency should consider imposing stricter limits on NO_x emissions from gasoline-fueled vehicles. Such actions will facilitate the agency's development of control strategies for diesel cars, because the control of NO_x emission from gasoline vehicles is easier than from diesels.

ENVIRONMENTAL EFFECTS

Smoke, noise, and odor are three perceptible nuisance factors associated with diesels. Of these the most significant is smoke, which contains particulates. Because particles of diesel exhaust are capable of absorbing visible light, greater numbers of diesels are likely to decrease the average visibility range, especially in certain urban areas.

Using carbon monoxide as a surrogate of particulate dilution rates, estimates have been made of the increase in ambient particulate concentrations in four cities with potentially large numbers of diesels. Estimates also have been made of the reduction in visibility in urban areas when the proportion of diesels in the entire fleet reaches 25 percent. In that event, the inhabitants of Los Angeles may experience as much as a 20 percent reduction in visibility, and in Denver a 50 percent decrease. The severest loss in visibility from a large increase in diesel cars is likely to occur in the "street canyons" of urban centers such as New York City. By contrast, in rural areas the decrease in visibility is expected to be less than 3 percent.

It is difficult to assess the effects of diesel emissions on photochemical reactions (atmospheric transformations in the presence of solar radiation), which may result in the formation of new chemical species, including aerosol components, with unexpected health and environmental consequences. Chemical and physical transformations in the atmosphere are not well enough understood to make even tentative conclusions. For this reason a major experimental program is needed to study the properties and reactions of diesel exhaust under simulated atmospheric conditions, using advanced analytic and smog chamber methods to study aerosol and reactive gas behavior.

HEALTH EFFECTS

After reviewing _in vitro_ tests, the committee's Health Effects Panel concluded that variations in fuel composition and engine operating characteristics may be significant factors in the biological activity of diesel exhaust components. Studies of _in vitro_ mammalian cell systems indicate that some diesel engines emit particulates containing sufficient levels of known mutagenic and carcinogenic compounds to produce cell transformations at high exposure levels.

However, in many instances the data provide few answers and raise many questions about the biological activity and potential carcinogenicity of diesel exhaust. The panel found no consistent pattern of effects in _in vitro_ studies of diesel emissions compared with samples of gasoline engine exhaust, coke oven emissions, roofing tar, and cigarette smoke. Even so, in limited comparisons with extracts of exhaust collected from gasoline engines, diesel exhaust extracts appear to contain more direct-acting mutagens. In other studies, diesel extracts, as well as emission extracts from comparable gasoline engines, induce skin cancer when applied to the backs of mice. But neither whole diesel nor whole gasoline engine emissions inhaled by various types of laboratory animals in different experiments has been found so far to induce lung tumors—due possibly to the low carcinogenic activity of the exhaust of both types of engines.

Present information suggests that pulmonary defense mechanisms may be adversely affected by some components of diesel exhaust—e.g., NO_x, which has been shown to decrease resistance to infectious diseases in both animals and humans. Additional information suggests that a single high-level exposure of diesel exhaust can produce toxic effects—e.g., poisoning due to NO_x and to aldehydes—whereas long-term exposure to comparatively low diesel exhaust levels has not clearly been shown to cause pulmonary and systemic toxicity. It is reasonable to expect that health hazards associated with certain pollutants originating in diesel exhaust (SO_x, NO_x, O_3, and possibly particulate material) would be qualitatively similar to those associated with the same pollutants in other sources—e.g., fossil-fueled power plants.

No convincing epidemiologic evidence exists in studies of groups exposed to diesel exhaust in the workplace that there is a connection between diesel fumes and human cancer. However, as there are only two studies that even approximate the minimum requirements for a sound epidemiologic evaluation and these suffer from flaws in their design, including a failure to account for cigarette smoking, the negative conclusions must be viewed with caution. Whether there is a relation between exposure to diesel exhaust and prevalance of chronic obstructive lung disease is uncertain. Some studies have suggested that workers exposed to diesel exhaust have a higher prevalance of chronic respiratory symptoms and bronchitis, as well as diminished lung function, than otherwise comparable persons who have not been exposed. Other studies have failed to confirm these observations.

More information is needed on the incidence of cancer, particularly lung cancer, and other non-malignant diseases such as chronic bronchitis and emphysema in persons who are heavily exposed to diesel emissions, comparing the results for persons in similar circumstances who are not exposed.

ECONOMIC EFFECTS

Few issues involving technology are without economic impacts. The issue of diesel cars concerns the motoring public, the automotive industry, and the whole society. To motorists, today's diesels have higher purchase prices and higher maintenance expenses but lower average total costs than comparable gasoline-powered cars. The advantages of the diesel may outweigh the disadvantages financially, provided the car is kept 10 years or more, runs at least 100,000 miles, and operates mostly in stop-and-go urban traffic where the diesel's relative fuel economy is greatest. Overall fuel savings of a large diesel car could amount to $1,800 over 10 years, assuming that the real price of motor fuels rises to $2 a gallon by 1990. For small diesel cars the savings may be around $950. The total savings, taking the purchase price and maintenance costs into account, could be $600 for a large diesel car and $360 for a small one in the 10 year period.

For manufacturers, the stakes in correctly projecting the diesel market are substantial. If diesels were to make up 18 percent of the new sales of light vehicles in 1990, approximately 3 million diesel passenger cars and small trucks would be produced in that year. If U.S. car makers built 80 percent of these, they would be producing about 2.5 million diesels in 1990, which would require a capital investment of $3 billion to $4 billion. Although a large sum, the investment must be viewed in perspective. It will be made over the period of a decade--some of it has already been made--and it will be a fraction of the total investment of U.S. automobile manufacturers for new plants, new designs that include microprocessor systems, and new tools such as computer-controlled robots.

Society at large may be the major beneficiary of "dieselization" if the demand for oil is reduced, because the market price of oil does not take into account its full cost in terms of the nation's vulnerability to political pressures and price rises imposed by foreign oil producers. Higher oil prices reduce purchasing power and contribute to inflation.

The availability of diesels may also confer a benefit to society in terms of road safety. Because diesels can achieve the same fuel economy as smaller gasoline-powered cars, dieselization makes it possible to attain any given level of average fleet fuel economy, as required by the corporate average fuel economy (CAFE) standards with a larger average car size. Dieselization also might reduce the number of traffic fatalities. In 1979, the most recent year for which complete statistics are available, small cars were involved in 55 percent of all fatal crashes, even though they accounted for only 38 percent of the cars on the roads, according to the National Highway Traffic Safety Administration. Based on traffic accident statistics for the 1960's and 1970's, a study performed for the committee indicates there might be about 800 fewer road deaths per year if 10 percent of the automobile fleet consisted of diesel cars that are larger and heavier than the same percentage of gasoline-fueled cars replaced. Still, changes in vehicle safety design, use of passive restraints, and improvements in gasoline-engine fuel economy may decrease the apparent safety disparity between large and small cars.

REGULATORY POLICY ANALYSIS

In the cost-benefit analysis of regulatory alternatives provided by the committee's Analytic Panel, the problems of poor and insufficient data as well as uncertainty are recognized. Two alternatives were selected for the analysis. The first is the current EPA plan for imposing the 0.6 g/mi particulate standard from 1982 through 1984 and the 0.2 g/mi standard thereafter. The second is the retention of the 0.6 g/mi particulate level beyond 1984.

Compliance with the 1982-1984 standard of 0.6 g/mi is technologically feasible at costs ranging from zero to $30 per vehicle. Therefore, the committee holds the EPA standard of 0.6 g/mi for diesel car particulate emissions to be both practical and prudent as a minimum level of safeguarding public health and the environment.

The panel's analysis takes account of the effect on the cost of purchasing and operating diesel cars with emission control devices to attain the 0.2 g/mi particulate standard. The cost of achieving the 0.2 g/mi standard, in contrast to holding to the 0.6 g/mi standard, are considerable--from $150 to $600 per vehicle. Under the 0.2 g/mi standard, diesel buyers will experience this cost increase as an economic loss. Some potential diesel buyers may be deterred by the cost increase and may choose gasoline-powered vehicles instead. If they choose gasoline-powered vehicles of equal size to the diesels they would otherwise have bought, their fuel consumption will be higher and the social benefit of reducing oil imports will be lost. If they choose gasoline-powered vehicles of equal fuel economy, the average car size will be smaller, and they may lose some amenities of larger cars, including the benefit of greater road safety. On the other hand, the 0.2 g/mi standard will reduce the total volume of diesel particulate emission, with consequent health and environmental benefits to the public at large.

The committee has identified the uncertainties relating to health effects of diesel particulates, the feasibility and cost of achieving the 0.2 g/mi standard, the determinants of behavior in the market for cars, the future conditions in the oil market, and the prospective regulatory policy approaches to automotive emissions. Although the committee has attempted to address the uncertainties explicitly, certain dimensions of the costs and benefits were omitted from its analysis and others were dealt with in summary ways. Thus, while cancer risks are considered, pulmonary and other systemic diseases are not. While automobile accident fatalities are included, non-fatal road injuries are not. Similarly, such environmental effects as soiling are not explicitly considered. The appropriate treatment of such uncertainties in the cost-benefit analysis may be complicated by incomplete or inaccurate data, ignorance about cause and effect, and value judgments, with questions of value perhaps the most serious and least tractable. In Chapter 7 the Analytic Panel explains the range of assumptions and values used in this analysis.

It is difficult, therefore, to reach definitive conclusions about the balance of costs and benefits in the comparison of the alternative particulate standards. The uncertainties could be resolved in ways favorable either to the 0.2 g/mi standard or the 0.6 g/mi standard. In the panel's analysis, neither policy is always dominant. Based on the current state of knowledge, an irrevocable decision by the EPA--whether it is 0.2 g/mi or 0.6 g/mi--could run a danger of costly mistakes.

To illustrate the costs of such mistakes, the Analytic Panel has estimated the losses to society that would result from the choice of either standard, assuming the uncertainties were resolved in a way favorable to the other. This analysis considers the "regrets" associated with a regulatory choice that could be wrong. Such an analysis depends on, among other things, the value assigned to preventing premature death from diesel emissions or traffic accidents. While the panel did not find a basis for adopting a specific assumption for the value of avoiding a premature death, it calculated the regrets associated with a range of values derived from studies of individual willingness-to-pay to avoid premature death. On this basis, the regrets associated with an incorrect choice of the 0.2 g/mi standard range from $2.0 billion to $2.9 billion per year. For an incorrect choice of the 0.6 g/mi standard, the costs range from $80 million to $1.2 billion per year. This suggests that a once-and-for-all choice under large uncertainties imposes substantial risks, but that the possible regret associated with the choice of the 0.2 g/mi standard is significantly greater than for retaining the 0.6 g/mi standard.

The use of regret analysis, the panel recognizes, is only one of many possible methods of decision-making. Such an analysis points to the policy that minimizes the worst outcome. The panel did not attempt to apply an alternative method. In particular, it did not assess the probabilities of each outcome, either subjectively or objectively, and then calculate the expected costs and benefits. Thus, the conclusions are especially sensitive to the extreme values chosen to represent the worst cases. Since the panel's assessment of the worst cases necessarily contains subjective elements, the specific numerical results should be viewed with caution.

CONCLUSIONS

The committee does not consider that commitment is warranted now
to the 0.2 g/mi standard for post-1984 diesel cars. Only a relatively
small number of diesel-powered cars will be made and sold in the next
few years, so the benefits and risks will appear slowly. Exposure to
diesel emissions, for instance, will be at a low level for some years,
with small risks to humans and the ecosystem. Meanwhile, within the
next two to five years, additional information and understanding of
health and environmental effects are likely to be forthcoming. In two
to five years, a more informed decision can be made about the impacts
of dieselization.

One of the motives for commitment to the future imposition of
stringent standards is to induce the development of more effective or
inexpensive emissions control technology. The committee concludes
that a more effective means to this end from the point of view of
benefits and costs to society is to establish procedures to formally
reevaluate the need for more stringent standards every three years,
beginning in 1983, and continuing for so long as the issue remains in
serious question. Such reevaluations should include new information
about soiling and health effects other than cancer, whose impacts
could not be quantified with the information available to the
committee. The approach is consistent with a sequential
decision-making process best suited to such uncertain situations. To
support such reassessments, more government-supported research is
warranted on the health and environmental effects of diesel emissions.

In addition to the two alternatives evaluated by the Analytic
Panel, several other regulatory approaches should be seriously
considered in future reassessments by the Congress and the EPA:

- Regulate particulate exhaust from such large sources of
  emissions in road transport as heavy diesel trucks and buses;
  this may be more cost-effective than tightening the emission
  levels of diesel cars and light trucks.
- Explore intermediate levels of the diesel particulate
  regulatory standard, between 0.2 g/mi and 0.6 g/mi.
- Develop corporate average particulate emission standards
  similar to the corporate average fuel economy (CAFE)
  standards now in effect, which would provide greater
  flexibility in attaining any given limit on total particulate
  emissions, by replacing the uniform standard applicable to
  each individual diesel vehicle.
- Institute state standards rather than uniform national levels
  to permit the application of stricter standards in air basins
  with particularly severe problems, as in the case of
  California air pollution standards for hydrocarbons and NO_x.
- Levy emission charges that vary with the quantity of emission
  rather than impose uniform mandatory standards; this will
  provide greater flexibility to manufacturers and car buyers
  and achieve any given limit on total pollution at minimum
  social cost.

CONTENTS

1 INTRODUCTION

In 1950 the Mercedes-Benz diesel was introduced in the United States as the first passenger car of its type. Only four were sold that year. Diesel engines were best known then to truckers, farmers, railroad engineers, and boat owners. By 1974, having experienced an oil embargo, a four-fold rise in the price of gasoline, and long lines at the filling pumps, Americans began buying diesel cars--some 11,000 Mercedes-Benzes and Peugeots, which accounted for 0.1 percent of all new cars that year. In 1979, when the revolution in Iran caused another temporary oil shortage and gasoline prices advanced rapidly, the sale of new diesels accelerated to 280,000--still only 2.2 percent of all new cars. In 1980 the market in diesel cars increased to 387,000 or 4.6 percent of the nation's new cars, with 320,000 of them bearing U.S. brand names and, significantly, many running with engines made in West Germany and Japan.*

As small as this market for light-duty diesels still is, the trend toward "dieselization" is unmistakable. There are now 13 marques for light-duty diesel vehicles, operating on a variety of 4, 5, 6, and 8 cylinder engines in displacements from 1.6 to nearly 6 liters, and with body styles from subcompacts to station wagons and pickup trucks.

Both manufacturers and motorists are turning to diesels because the engines provide 25 to 35 percent greater average fuel savings than gasoline-powered cars and light trucks of comparable size. Thus, the expansion of light-duty diesels should help reduce the use of petroleum, which is a national political and economic goal. Assuming that diesels average 35 percent greater fuel efficiency than gasoline-powered vehicles, when diesels make up 20 percent of the passenger car fleet some time in the future, the use of oil will be reduced by 5 percent. In addition, the fuel economy of diesel engines is important for the car makers in meeting the U.S. Department of Transportation's corporate average fuel economy (CAFE) standard, which requires a level 27.5 mpg by the model year 1985.

Many car makers see the diesel as a way of marketing six-passenger family automobiles while meeting the 27.5 mpg CAFE standard. According to this strategy, as much as 10 percent of the passenger cars and light trucks sold in 1985 could be operating with diesel engines and perhaps 25 percent of the new cars in 1990 could be diesels.

---

*Sales data obtained from the Motor Vehicle Manufacturers Association.

1

But there is a roadblock to the propective dieselization of the nation's roads. Despite their significant fuel economy and the additional advantage of meeting current emission standards for hydrocarbons (HC) and carbon monoxide (CO) under the Clean Air Act, diesel cars produce, even when properly tuned and operated, more particulate matter and greater quantities of nitrogen oxides (NO_x) than catalyst-equipped cars running on unleaded gasoline. The particulates are of critical concern. They are formed from bits of aggregated carbon during incomplete combustion of diesel fuel, as well as from some components of lubricating oil. As soot particles they have a mean diameter around 0.2 μm, which means they can be inhaled deep within the lungs to penetrate the tracheobronchial and alveolar regions. There they can remain for weeks or years and aggravate chronic lung disease by disturbing normal ventilation and causing a reflex constriction of blood vessels. Extracts of diesel exhaust are known to contain chemical substances that are potential carcinogens.

The size and distribution of the particulates are important for other reasons. Suspended in the air, they can be responsible in some urban areas for reducing visibility and for catalyzing the formation of photochemical smog. Moreover, the high incidence of fine particles in diesel soot also contributes to smoke and odor, which many people find objectionable, as well as to soiling or staining of buildings and other structures.

By 1990, according to an analysis of the problem by the U.S. Environmental Protection Agency (EPA), light-duty diesel vehicles are "projected to become the seventh largest source of particulate emissions and to have the third greatest available potential for total particulate emission reduction of any source, mobile or stationary" (U.S. EPA, 1980). Because of the possible hazard to people and places, the EPA is required by Congress, under the 1977 amendments to the Clean Air Act, to regulate diesel exhaust. Thus, early in 1980 the agency promulgated the emission standard for particulate matter from diesel-powered cars and small trucks at 0.6 g/mi (grams per mile) beginning with the 1982 model year and tightened the standard to 0.2 g/mi for cars and 0.26 g/mi for light trucks beginning with the 1985 model year.

EPA set the 0.2 g/mi standard confident that a particulate control technology--namely, the trap oxidizer--would be successfully developed and commercially available for the 1985 cars. Why was a regulatory rule needed if an emission control device would be available? EPA provided a blunt answer: "Experience has shown that in the absence of direct regulatory incentive, manufacturers have rarely invested the necessary resources into new emission control technologies" (U.S. EPA, 1980).

Without emission control, EPA had calculated, light-duty diesels would emit between 152,000 and 253,000 metric tons of particulate matter by 1990, when sales of such vehicles reach 15 to 25 percent of all new cars and small trucks. EPA expects that the regulation will reduce particulate emissions from such vehicles by 74 percent in 1990 to 40,000 to 66,000 metric tons per year (U.S. EPA, 1980).

In addition to the particulate standard, the diesel car makers will have to comply with the 1.0 g/mi NO_x standard in the model year 1985 and thereafter. The history of both the particulate and NO_x standards for diesels is one of deadlines set and delays granted, usually for two-year periods, when the manufacturers assert they cannot meet the EPA timetable. The question of NO_x and particulate standards for diesel cars is a major issue before the EPA and probably no less important than the SO_x emission standard issued in 1979 for new coal-burning electricity plants.

Given the state of auto emission control technology then and now, a tightening of the NO_x and particulate standards to the prospective 1985 levels could frustrate the plans of car makers to advance dieselization.* This is likely to affect an industry that occupies a central position in the economy. The production of cars and trucks and their insurance, repairs, and servicing (excluding fuels) account for about 8.5 percent of GNP (U.S. Department of Commerce, 1981). The industry's size, complexity, and pervasiveness make it peculiarly sensitive to business conditions, consumer psychology, and government actions.

If the use of diesels is to expand considerably, manufacturers will have to make large investments in facilities during this decade. At the same time, government decisions that affect the expansion of diesels will also be made. At least at the outset, both sets of decisions are usually based on limited information and great uncertainty, though as new knowledge becomes available, decisions can be changed. If, for example, the initial decisions were to facilitate an unlimited expansion of diesels and a decade later diesel particulates were discovered to be a health hazard, the decisions could be reversed. By then, though, many people would have been exposed to the hazard, millions of diesels would be in use (and it is unlikely they would be ordered off the roads), and diesel production facilities would have to be closed or converted. In the opposite case, if diesels were prohibited at the start because of health risks, and a decade later diesel exhaust particulates were found to have negligible adverse effects on health, society would have been deprived of an important source of energy conservation, consumer satisfaction, competitive advantage, and regulatory credibility.

In all this the costs of reducing pollution are usually obvious and reasonably measurable, though the benefits are in large part intangible and unmeasurable. Even the social costs of improving the quality of air or reducing the risk of cancer are, if not fully known, at least knowable. The social benefits, however, are to a great extent in the realm of values and hence varied and speculative. Still, both costs and benefits need to be considered in regulatory actions, especially, as the British economist A.C. Pigou observed in the 1920's, when the marketplace cannot offer incentives for reducing the levels of pollution and the government is required to regulate to maintain public health and environmental quality.

---

*Of the 104,564,000 passenger cars on the nation's roads at the end of 1980, about 1 million were diesel-powered. A larger proportion of trucks have diesel engines. With 33,350,000 registered trucks of all weights and sizes operating in 1979, those equipped with diesel engines totaled 1,562,000, a 7 percent gain over the previous year. Of 520,370 buses in the nation, about 93,000 are diesels, a 6 percent increase in one year.

THE CLEAN AIR ACT AND AUTOMOTIVE VEHICLES

The Clean Air Act as amended in 1977 (PL 95-95) calls for a
two-stage procedure for pollution control. First, it directs the EPA
to establish primary and secondary national ambient air quality
standards for pollutants. The primary standards are set to "protect
the public health ... allowing an adequate margin of safety" [Sec.
109(b)(1)]. The secondary standards are to "protect the public welfare
from any known or anticipated adverse effects associated with the
presence of such air pollutant in the ambient air" [Sec. 109(b)(2)].

The primary standards are the immediate goals for air pollution
control effects; the secondary standards set the long-run goals for air
pollution control by providing more stringent requirements for lower
levels of pollutants in the air. Thus far, EPA has set standards for
carbon monoxide (CO), hydrocarbons (HC), nitrogen oxides (NO_x),
sulfur oxides (SO_x), particulates, photochemical oxidants, and, most
recently, airborne lead. In all cases, the standards have been
expressed as maximum permissible concentrations of the pollutant in the
ambient air (either parts per million or grams per cubic meter), based
on an average for some specific time period.

The second aspect of the act is directed at reducing emissions of
the pollutants from stationary and mobile sources in order to achieve
ambient air quality standards. The reductions apply to three sources:
existing stationary sources, through implementation plans that are
developed by the states and approved by EPA; new stationary sources,
such as new factories or electric power plants, through "new source
performance standards" established by EPA; and new motor vehicles,
through standards either set directly by the act or by EPA in
accordance with the provision of the act. Among the states, California
alone is allowed to set more stringent standards for motor vehicles
because of the severe pollution problems in the Los Angeles basin.

EPA first began setting emission standards for motor vehicles for
the 1968 model cars. The current standards for automobiles and
light-duty trucks are provided in Tables 1.1 and 1.2. Current
hydrocarbon and CO emission limits for passenger cars represent roughly
a 90 percent reduction from the levels typically emitted by cars
without emission controls. The NO_x emission limit represents a
reduction of approximately 75 percent from that of typical uncontrolled
cars. Current light truck emissions standards represent about a 75
percent reduction from uncontrolled levels for hydrocarbons and CO and
a 60 percent reduction in NO_x. The act requires a 90 percent
reduction from uncontrolled levels for light trucks by the 1983 model
year for hydrocarbon and CO and a 75 percent reduction by the 1985
model year for NO_x.

Moreover, the act mandates that particulate emission standards be
set for all classes of vehicles beginning with the 1981 model year.
The regulations need to reflect the greatest degree of emission
reduction achievable with the available technology, considering the
cost of applying the technology, as well as noise, energy, and safety
factors [Sec. 202(a)(3)(A)(iii)]. Particulate standards for
automobiles and light trucks were established in February 1980.
Maximum emissions for the 1982-1984 models are set at 0.6 g/mi for both

TABLE 1.1   Gaseous Emission Standards for Light-duty Vehicles
Grams Per Mile (grams per kilometer).

Model Year	HC	CO	NO$_x$	HC Evap.[a]
		Federal		
1978-79	1.5 (0.93)	15.0 (9.3)	2.0 (1.24)	6.0
1980	0.41 (0.25)	7.0 (4.3)	2.0 (1.24)	6.0
1981 and on	0.41 (0.25)	3.4 (2.1)[e]	1.0 (0.62)[d]	2.0
		California		
1978-79	0.41 (0.25)	9.0 (5.6)	1.5 (0.93)	6.0
1980[c]	0.41 (0.25)	9.0 (5.6)	1.0 (0.62)	2.0
1981-A[b,c]	0.41 (0.25)	3.4 (2.1)	1.0 (0.62)	2.0
1981-B	0.39/0.41 (0.24/0.25)	7.0 (4.3)	0.7 (0.43)	2.0
1982-A[b,c]	0.39/0.41 (0.24/0.25)	7.0 (4.3)	0.4 (0.25)	2.0
1982-B	0.39/0.41 (0.24/0.25)	7.0 (4.3)	0.7 (0.43)	2.0
1983[c] and on	0.39/0.41 (0.24/0.25)	7.0 (4.3)	0.4 (0.25)	2.0

[a]   SHED test (grams per test).

[b]   Manufacturers have the option of using "A" for 1981 and 1982 or of
using "B" for 1981 and 1982. Also, manufacturers have a choice
between a 0.24 g/km non-methane hydrocarbon standard and the 0.25
g/km total hydrocarbon standard.

[c]   If emission durability is established for 160,000 km (100,000
miles) the NO$_x$ standards for option A are 0.93 g/km (1980-81) and
0.62 g/km (1982-83).

[d]   Waiver for diesels to 1.5 g/mi (0.93 g/km) possible for 1981-1984
model years.

[e]   Waiver to 7.0 g/mi possible for 1981-1982 model years.

TABLE 1.2   Gaseous Emission Standards for Light-duty Trucks
Grams Per Mile (grams per kilometer).

Model Year	HC	CO	NO$_x$	HC Evap.[a]
Federal				
1978	2.00 (1.24)	20.0 (12.4)	3.1 (1.93)	6.0
1979[b]	1.70 (1.06)	18.0 (11.2)	2.3 (1.43)	6.0
1980 and on	1.70 (1.06)	18.0 (11.2)	2.3 (1.43)	6.0
1981 and on				2.0
1983 and on	to be determined			
California (0–5999 GVWR)				
1978	0.90 (0.56)	17.0 (10.6)	2.0 (1.24)	6.0
1979–80 (0–3999 IW)	0.41 (0.25)	9.0 (5.6)	1.5 (0.93)	
(4–5999 IW)	0.50 (0.31)	9.0 (5.6)	2.0 (1.24)	
1979				6.0
1980 and on				2.0
1981–82 (0–3999 IW)	0.41 (0.25)	9.0 (5.6)	1.0 (0.62)[c]	
(4–5999 IW)	0.50 (0.31)	9.0 (5.6)	2.0 (1.24)[d]	
1983 and on				
(0–3999 IW)	0.41 (0.25)	9.0 (5.6)	0.4 (0.25)[e]	
(4–5999 IW)	0.50 (0.31)	9.0 (5.6)	1.0 (0.62)[c]	
California (6,000 and Larger GVWR)				
1978–80	0.90 (0.56)	17.0 (10.6)	2.3 (1.43)	
1978–79				6.0
1980 and on				2.0
1981–82	0.60 (0.37)	9.0 (5.6)	2.0 (1.24)[f]	
1983 and on	0.60 (0.37)	9.0 (5.6)	1.5 (0.93)[d]	

a   SHED test (grams per test).

b   Federal weight class for LDT charges from 0–6000 GVWR to 0–8500 pounds CVWR.

c,d,e,f If emission durability is established for 160,900 kilometers the NO$_x$ standard is: 0.93,[c] 1.24[d] 0.62,[e] or 1.43[f] grams per kilometer.

kinds of vehicles; maximum emissions for 1985 and after are 0.2 g/mi for automobiles and 0.26 g/mi for light trucks. Separate standards are currently being formulated by EPA for heavy trucks.

While diesel vehicles can meet current HC and CO emission standards under the Act, they are hard put to meet the proposed maximum level for particulates or soot. The emissions limits are designed primarily to assist communities in meeting the national ambient air quality standard for particulates.

The possible carcinogenic properties of diesel particulates are not addressed by the standards. If EPA were to conclude that diesel particulates pose a threat more serious than the health hazard of ambient particulates, it could issue regulations to curtail the emissions under the authority of Section 202(a)(1) or refuse to certify the vehicle for sale under sections 202(a)(4) and 202(a)(3).

Thus far, the vehicle emissions limits for HC, CO, NO_x, and particulates have been based on the precept that every vehicle must meet the standards for its first five years or 50,000 miles. In practice, this has worked out as follows. Pre-production models of each vehicle undergo a specified set of emissions tests. Each vehicle in a small group of models is run for 50,000 miles and its emissions tested every 5,000 miles. A deterioration factor, based on a least-squares regression of the periodic emissions readings and mileages, is developed for each vehicle. Then a larger family of vehicles is tested for only 4,000 miles, with the 50,000-mile emissions predicted from the deterioration factors. If a vehicle's emissions remain under the EPA limits throughout the 50,000 miles (either by actual observations or extrapolations), the vehicle is certified for production. If the vehicle fails, the manufacturer needs to start the test series over with the same (or a modified) model.

EPA has been making assembly line checks of vehicles since 1976. Currently, only 60 percent of the assembly line cars are expected to pass the test, but EPA has recently established a 90 percent pass rate for assembly line checks of heavy-duty trucks, and it is possible the agency will eventually require that 90 percent of the cars and light-duty trucks pass the emissions tests. The 90 percent pass rate is considered by EPA to be equivalent to requiring every vehicle to pass the standard. EPA can levy fines of up to $10,000 for each vehicle that does not pass the test but is sold nevertheless.

Despite EPA's pre-production tests and assembly line tests, cars in actual use have been deteriorating faster in their emissions performance, on average, than the tests predict. For a number of years, EPA has been testing a sample of cars on the road, using the same certification test given to the pre-production models. Table 1.3 shows the data for tests conducted in 1978. Over a third and in some cases well over half of the in-use vehicles tested exceeded the applicable emissions limits for HC and CO. Their performance with respect to NO_x, though, is considerably better.

FUEL ECONOMY REGULATIONS

The Energy Policy and Conservation Act of 1975 provides the basic structure for fuel economy regulation. The National Highway Traffic

TABLE 1.3 Average Pollution Emissions from Motor Cars in Use in Six Cities[a] in 1978.

Model Year	No. of Vehicles	Avg. Mileage	Avg. HC Emissions (g/mi)	% Above 1.5 g/mi	Avg. CO Emissions (g/mi)	% Above 15 g/mi	Avg. NO_x Emissions (g/mi)	% Above 3.1 g/mi
1975	262	38,010	2.30	60	32.93	65	2.50	25
1976	413	27,464	2.25	57	29.50	54	2.36	17
1977	636	16,478	1.61	38	23.49	51	1.88	27[c]
1978[b]	48	7,386	1.86	46	29.72	58	1.31	4[c]

[a] Chicago, Denver, Houston, Phoenix, St. Louis, and Washington, D.C.

[b] St. Louis and Denver only.

[c] Percent above 2.0 g/mi.

Safety Administration (NHTSA) has the major regulatory responsibility. The act stipulates specific fuel economy standards for automobiles for the 1978-1980 and 1985 model years. NHTSA has specified the standards applicable to the 1981-1984 model years. The standards are listed in Table 1.4. For 1985 and after, NHTSA is empowered to raise the fuel economy standards to a maximum of 29.0 mpg. Any higher level requires new legislation. NHTSA's standards for light-duty trucks through 1981, are also listed in Table 1.4.

Fuel economy standards apply to the sales-weighted average of each manufacturer for the applicable class of vehicle. Thus, the fuel economy of individual car models can fall below the standard so long as the sales-weighted average for the manufacturer is at or above the applicable standard. Manufacturers are allowed to carry forward or carry back any sales-weighted margins above or below the standard for three years. Failure to meet the standard is considered a violation of the law, subjecting the manufacturer to a fine of $5 per vehicle for each 0.1 mpg for the entire number of vehicles sold in that class.

There is a second, less well known, piece of legislation that also affects fuel economy. The Energy Act of 1978 imposes a "gas guzzler" tax on individual automobile models that fall below specified fuel economy levels. This tax is entirely separate from the fuel economy standards set by the Energy Policy and Conservation Act. The schedule of taxes for 1981 vehicles ranges from $200 for 16 to 17 mpg to $650 for less than 13 mpg. The range of miles per gallon narrows and the rate of taxes increases progressively through 1986, when they will be $500 for 21.5 to 22.4 mpg and $3,850 for less than 12.5 mpg.

Finally, one other influence on fuel economy is the level of taxes on fuel. Higher taxes may induce consumers to choose more fuel-efficient vehicles. Currently, federal and state taxes on gasoline and diesel are about 12 cents per gallon, with a federal component of 4 cents. The taxes are largely designated for highway construction and maintenance, so that the taxes could be considered, roughly, highway user fees. Nevertheless, it is possible to think of highway expenditures being financed by other sources of taxation. In this sense, the 12 cents of taxes raises the price of fuel and therefore helps to encourage fuel conservation and promote fuel economy.

SUMMARY OF CURRENT POLICY INSTRUMENTS

The policy instruments used to regulate emissions and fuel economy described here encompass a fairly broad spectrum of approaches. The approaches can be organized into categories:

- Emission standards that every car must pass and severe penalties--tantamount to a ban on production--if models fail to meet the levels set by EPA;
- Fuel economy standards set by NHTSA, with fleet averaging permitted and modest penalties imposed for violators;

TABLE 1.4 Fuel Economy Standards for Automobiles and Light-Duty Trucks.

| Model Year | Automobile | Light-duty Trucks | |
		2-Wheel Drive	4-Wheel Drive
1978	18.0[a]		
1979	19.0[a]	17.2[b] [c]	15.8[b] [c]
1980	20.0[a]	16.0[b] [c]	14.0[b] [c]
1981	22.0[b]	16.7[b] [c]	15.0[b] [c]
1982	24.0[b]		
1983	26.0[b]		
1984	27.0[b]		
1985	27.5[a]		

[a] Mandated by the Energy Policy and Conservation Act of 1975.

[b] Established by NHTSA rulemaking.

[c] 1979 standard applied to vehicle up to 6,000 lbs; 1980 and 1981 standard applied to vehicles up to 8,500 lbs.

- Emission standards that apply only to particular localities, as is the case with the stricter California standards;
- Taxes on vehicles according to a schedule of achievement, as is the case with the "gas guzzler" tax; and
- Federal and state taxes on motor fuels.

The first four classes are ways of regulating either emissions or fuel economy; the last class is limited strictly to fuel economy.

REFERENCES

U.S. Department of Commerce (1981). Bureau of Economic Analysis (unpublished data).

U.S. Environmental Protection Agency (1980). Regulatory Analysis of the Light-Duty Diesel Particulate Regulations for 1982 and Later Model Year Light-Duty Diesel Vehicles. Office of Mobile Source Air Pollution Control. Washington, D.C.

2 DIESEL TECHNOLOGY AND EMISSIONS

In the history of internal-combustion engines, two names predominate--Nicolaus August Otto and Rudolph Christian Karl Diesel. Otto's four-stroke spark-ignition engine, based on the concepts of Alphonse Beau de Rochas, received a patent in 1876--sixteen years before Diesel's patent for a two-stroke compression-ignition engine.

Diesel's engine is similar to Otto's, but does not depend on an electric spark for ignition of the mixture of fuel and air. In the diesel, a charge of air is drawn into a cylinder on the down stroke of the piston and is compressed on the up stroke to a temperature hot enough to ignite the fuel without a spark when it is sprayed into the cylinder. Diesel conceived his engine as an improvement on the thermal efficiency and operating cost of the Otto cycle by applying the ideas set forth by Nicolas Leonard Sadi Carnot in 1824. Carnot, for his part, had sought to increase the efficiency of James Watt's steam engine through temperature differences in the cylinders.

The diesel did not come to prominence as rapidly as Otto's engine. This was so because the first diesels were well beyond the state of the art at the time and, then, the early diesels were large and heavy in proportion to their power output and operated sluggishly. Lighter weight and more powerful, Otto's engine was quickly adopted and developed by Gottleib Daimler, Wilhelm Maybeck, and Karl Benz for their pioneering automobiles, thus championing its utility and universality. By contrast, it took Maschinenfabrik Augsburg A.G. five years to develop the diesel, and, in 1898, Adolphus Busch, a beer brewer in St. Louis, bought the rights from Diesel to build and sell the engines in the United States.

Diesels can use heavier fractions of petroleum than Otto engines. This means that diesel fuel, being relatively unrefined, is cheaper than gasoline and, being less flammable, is safer. Moreover, diesel fuel has a larger energy content than gasoline and, because diesels burn it more efficiently, the engines are more appealing economically. Still, when Henry Ford was making his first passenger cars, he rejected the large and heavy diesel in favor of the Otto engine. Nor did the Wright brothers find it usable for their airplanes.

Not until World War I did diesels find an application--as the power plant for German submarines. The engines soon proved admirable for ships and locomotives, so that oil began to replace coal in heavy transport vehicles between the two world wars. It was not until 1923, however, that a practical diesel engine was produced in Germany for a truck. Meanwhile, in the United States, Clessie L. Cummins, founder of the Cummins Engine Co., foresaw the future of the diesel and undertook its development. In 1931 he set a new record at Daytona, Florida, by driving a diesel-powered Dusenberg at 100.75 mph. The 1934 Indianapolis "500" race had two Cummins diesel-powered cars--one a four-stroke engine and the other a two-stroke design. The four-stroke engine went on to Daytona to establish a new speed record of 137 mph.

Why did it take nearly another half century after Cummins's successes for the U.S. automobile industry to introduce a diesel-powered production model of passenger car--the General Motors Oldsmobile 5.7 liter diesel in the fall of 1977? The answer rests with the diesel's prolonged period of development* and with some of its characteristics--high noise level, thick smoke, acrid odor, poor cold weather starting, and torpid acceleration. These outweighed the diesel's fuel economy so long as U.S. gasoline prices were low. Thus, on the eve of the oil embargo imposed by the Arab members of the Organization of Petroleum Exporting Countries in 1973, U.S. gasoline prices averaged less than 30 cents a gallon. Not surprisingly, the nation's motor cars at that time averaged only 13.9 mpg.

IN-USE CHARACTERISTICS OF DIESEL ENGINES

Both the direct-injection and indirect-injection diesel engine for light-duty vehicles are currently less developed than the familiar gasoline engine, which has undergone many improvements in the past 20 years. Not unexpectedly, trouble appeared in the fuel injection system of some early models of light diesels. Water separators were inadequately designed and lacked dashboard warning lights to alert drivers of water in the fuel. (Moreover, excessive water was found in the fuel storage tanks of some local fuel distributors.) Some diesels have needed more frequent changes in lubricating oil and some have experienced unexpected wear to the valve train. Structural problems, normally encountered when new products are introduced, have appeared. These problems can be solved by improved designs, materials, and operational practices. There are no fundamental reasons why diesel engines cannot be as durable, reliable, and serviceable as gasoline engines.

To achieve trouble-free high performance, diesels need further improvements in the following areas: fuel injection systems, electronic engine controls and sensors, exhaust gas recirculation, blowers or turbochargers, transmissions and transmission control, lubricant control, oil additives that are not affected by particulates, and regenerative trap-oxidizers or some other emission control device.

*For a brief history of the troubled evolution of Diesel's engine, see Lynwood Bryant (July 1976). "The Development of the Diesel Engine." Technology and Culture, Vol. 17, No. 3.

FUEL ECONOMY

Because of its higher density, primarily, a gallon of diesel fuel has approximately 13 percent greater energy (calorific) content than a gallon of gasoline. This difference is a main advantage of using diesel fuel. Comparisons on a mpg basis must take into consideration the present and future prices, including any price difference between gasoline and diesel fuel to the consumer, because the economics of dieselization is sensitive both to pump prices and to total annual mileage driven. On the basis of comparable performance, diesel engines have an approximate 35 percent mpg advantage over conventional gasoline engines in typical passenger cars. On an energy basis the advantage is approximately 20 percent. This figure varies however, depending on driving conditions and vehicle use. In city stop-and-go driving under light load, diesels may save 50 percent of the fuel consumed by a comparable gasoline-powered vehicle. The initial start-up and running characteristics of diesel engines (as opposed to the choke operation of spark-ignition engines) contribute to this fuel economy advantage. Driven on highways at high speeds, with heavy loads, diesels have an advantage of perhaps 15 percent.

Various vehicles and engine types are compared in Table 2.1. The table shows that the fuel economy advantages of the diesel over the gasoline engine are consistent and substantial for all makes of passenger cars, station wagons, and small trucks.

Test Methods and Data

Vehicle efficiency is generally expressed in terms of a ratio of fuel consumed to distance traveled (miles per gallon, litres per 100 km, km per liter). Most fuel economy data for the various engine/vehicle configurations come from vehicle certification tests conducted by the EPA. Such data are the result of chassis dynamometer testing (vehicle testing in laboratories) over specified driving cycles. The Federal Test Procedure that is considered to be typical of urban driving is used in vehicle certification tests. Whether chassis dynamometer testing provides accurate data for fuel economy has been a subject of controversy. Critics of the testing procedure claim that track, road, and in-use testing provide better data. In addition to this, the automotive industry has stated that the Federal Test Procedure cycle understates in-use fuel economy. In response, the EPA added a higher speed cycle, the Fuel Economy Test. A weighted combination of the Fuel Test Procedure (for urban motoring) and Fuel Economy Test (for highway driving) figures is used in determining compliance with the Corporate Average Fuel Economy(CAFE) requirements. Several comparisons between the EPA test procedures and actual road use tests, based usually on the fleet, have been conducted. These indicate that the Federal Test Procedure fuel economy number is probably closer to actual fuel use in city driving, though some biases are evident. More importantly, the test results show that the fuel economy advantage of diesel vehicles over gasoline vehicles is systematically

TABLE 2.1 Fuel Economy Data for Light-duty Vehicles, 1981 Model Year, EPA FTP (Urban) Cycle.

Vehicle Class	Vehicle Make and Model	Estimated mpg	Engine CID	Trans-mission	Fuel System No. bbls. or Fuel Injection	Fuel Economy Advantage (mpg)
Subcompact Cars	VW Rabbit	28	105	M4	Fl	
	VW Rabbit Diesel	42	97	M4	Fl	50%
Compact Cars	Audi 5000	19	131	M5	Fl	
	Audi 5000 Diesel	27	121	M5	Fl	42%
	Peugeot 505*	19	120	A3	Fl	
	Peugeot 505*	16	120	M5	Fl	
	Peugeot 505 SD	29	141	M&A4	Fl	53-81%
	Mercedes 280E*	16	168	A4	--	
	Mercedes 240D Diesel*	24	183	A4	Fl	50%
Mid-Size Cars	Oldsmobile Cutlass	21	231	A3	2	
	Oldsmobile Cutlass	19	260	A3	2	
	Oldsmobile Cutlass Diesel	23	350	A3	Fl	10-21%
Large Cars	Buick LeSabre	19	231	A3	2	
	Buick LeSabre	18	252	A3	4	
	Buick LeSabre	16	307	A3	4	
	Buick LeSabre Diesel	22	350	A3	Fl	16-38%
	Cadillac deVille/Brougham	18	252	A4	Fl	
	Cadillac deVille/Brougham Diesel	21	350	A3	Fl	17%

Mid-Size Station Wagons	Oldsmobile Cutlass	21	231	A3	2
	Oldsmobile Cutlass	17	260	A3	2
	Oldsmobile Cutlass	16	307	A3	4
	Oldsmobile Cutlass Diesel	23	350	A3	F1	10-44%
Large Station Wagons	Pontiac Catlina/ Bonneville Safari	16	307	A4	F1
	Pontiac Catalina/ Bonneville Safari Diesel	21	350	A3	F1	31%
Standard Pickup Trucks	Chevrolet C10	17	250	A3	2
	Chevrolet C10	17	305	A3	2
	Chevrolet C10 Diesel	20	350	A3	F1	18%

* 1981 certification data are not available. Data are for the 1980 model year.

(Source: EPA 1980, 1981)

underestimated by about 5 percent in the Federal Test Procedure. This is important in any comparison of the EPA listing of vehicle mileage and fuel economy ratings for gasoline and diesel vehicles.

An additional advantage of the diesel is its improved fuel economy on start-up and on short trips of less than 10 miles. Because of the high percentage of trips under 10 miles, this factor may represent a significant energy savings for a large portion of motorists in the United States. The diesel achieves fuel saving because it reaches normal operating conditions quickly and does not use overall fuel-rich mixtures during initial operation.

The city, highway, and combined ("55/45 mpg") fuel economies of EPA certification vehicles, sales-weighted for the passenger car fleet, are displayed in Figure 2.1. The in-use fuel economy ("road mpg") and the average inertia weight of vehicles, which indicates the downsizing of motor cars that has taken place since 1976, are also depicted. Figure 2.2 shows the fleet fuel economy normalized to the 1978 mix of inertia weight. The data indicate that improvements in fuel economy and reductions in tailpipe emissions were obtained simultaneously in 1975 (Murrell, 1980). Since 1977, downsizing has been the principal factor in better gasoline mileage for passenger cars.

Projections

Table 2.2 shows in summary form the fuel economies of today's light-duty vehicles, with both gasoline and indirect-injection diesel engines, compared with engines of the future--direct-injection, stratified-charge, direct-injection diesel, improved gasoline engines, and advanced-ignition diesels, which are all projected for production in the next two decades.

In the next five to ten years, increasing sales of direct-injection diesel passenger cars and small trucks appear to be likely because of their fuel economy and structural durability. At the same time, improvements in gasoline engines are possible, particularly through the use of "lean-burn" fuel-to-air ratios and stratified-charge concepts. Increasingly stringent regulation of exhaust emissions may affect the development and use of gasoline engines just as they may affect the development and use of diesels. For instance, the technology of reduction catalyst control of NO_x is currently not adaptable to the stratified-charge, lean-burn, or diesel engines because they all operate with an overall fuel-lean (oxidizing) mixture.

Other light-duty vehicle power plants are either available or in design and development. The only one that seems likely to be widely applicable for passenger cars and small trucks during this century is the battery-powered electric motor. The potential efficiency advantage of Stirling and Brayton (gas turbine) cycle engines ensures that they will be of enduring interest and will undergo continuing development, but it is unlikely that either of these engines will have an important impact on light-duty automotive applications during the next 20 years.

Figure 2.1 Sales-Weighted Fleet Fuel Economy Trends

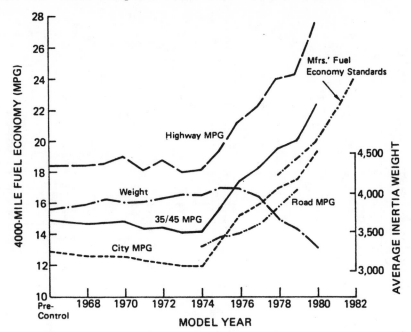

Source: Murrell, 1980

Figure 2.2 Weight-normalized Automobile Fuel Economy Trends

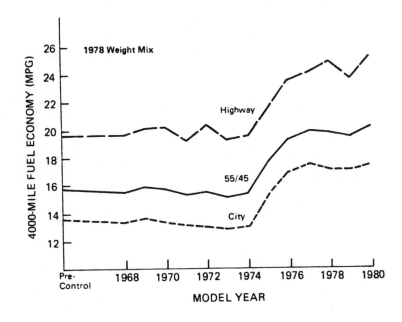

TABLE 2.2   Fuel-Economy Comparison of Light-duty Vehicles:
Comparison at Constant Performance in Percent with Respect
to Current Gasoline Engine Vehicles.*

PERIOD	ENGINE TYPE	VOLUME BASIS (MPG)	ENERGY BASIS
Current (1980)	Gasoline	0	0
	Indirect Injection Diesel	+35	+19
Intermediate (1980's)	Improved Gasoline	+20	+20
	Direct Injection Stratified Charge Gasoline	+25	+25
	Improved Indirect Injection Diesel	+45	+28
	Direct Injection Diesel	+55	+37
Long Range (>1990)	Advanced Stratified Charge Gasoline	+42	+42
	Advanced Ignition Assisted Diesel	+60	+42
	Advanced Fuel-Tolerant Engine	+40 to +65	+42

*Estimated uncertainties to any absolute value are $\pm$ 10 percent of value.
Difference between volume basis and energy basis are based on +13 percent
greater volumetric energy content of diesel fuel.

Particulates

The formation of particulates is fundamental to the combustion process in diesel engines, but the emission of particles in diesel exhaust is not. Small diesels emit from 30 to 100 times more particulate matter than comparable spark-ignition engines in vehicles fitted with catalytic emission controls. In the 1979 model year, light-duty vehicles with gasoline engines emitted particulates at an average of 0.02 g/mi, according to EPA data. For the same model year, particulate emissions from light-duty diesels ranged from 0.23 to 0.84 g/mi.

The comparatively high particulate level is the principal emission disadvantage of the diesel engine. Particulate formation begins when fuel is injected into the combustion chamber and continues during and after the dilution of the exhaust in the atmosphere. The first steps of the process occur rapidly--within microseconds or milliseconds. The last steps may take hours or even days.

Particulates emitted in diesel exhaust are the result of both formation and subsequent oxidation processes. Experimental evidence from a variety of combustion systems (premixed and diffusion flames, perfectly stirred reactors, etc.), as well as fuels, indicates that chemical kinetics is the dominant factor governing the formation and oxidation of particulate matter. Well-mixed systems emit particulates when the carbon-to-oxygen molar ratio in the fuel-oxidizer mixture exceeds 0.5 (Amann et al., 1980). The diesel has neither a premixed nor homogenous system.

The particles are carbonaceous solid-chain aggregates on which organic compounds are adsorbed. Most particles are the result of incomplete combustion of fuel hydrocarbons. Some come from the lubricating oil. The heaviest portions of the extractables from particulates have properties very similar to engine lubricating oil.

At combustion temperatures above 500°C, the small agglomerated chains of particles are composed principally of carbon-hydrogen spheres with diameters ranging from 100 to 800 angstroms (Å) (Vuk et al., 1976; Carpenter and Johnson, 1979). As temperatures decrease below 500°C, however, the resulting particles become coated with adsorbed and condensed species in the form of high molecular weight organic compounds, including:

- Unburned hydrocarbons;
- Oxygenated hydrocarbons (ketones, esters, ethers, organic acids) polynuclear aromatic hydrocarbons; and
- Inorganic species such as sulfur dioxide, nitrogen dioxide, and sulfuric acid (sulfate) (Khatri et al., 1978; Vuk et al., 1976).

Diesel particulate is generally defined as any material that is collected, at a temperature of 52°C or less, on a filtering medium after dilution of the raw exhaust gases.

The size of diesel automobile particulates peaks at about 0.15 μm, while catalyst-equipped gasoline engines emit particulates with

peaks at 0.02 μm. Such dimensions are somewhat misleading because
they represent equivalent spherical aerodynamic diameters, but diesel
particles occur as agglomerated chains. Photomicrographs show that
diesel particulates have a very light, fluffy structure, with a density
of about 0.07 g/cm^3 (Vuk et al., 1976; Khatri et al., 1978; Carpenter
and Johnson, 1979). Because of their properties, diesel particulates
tend to load up and plug any static filtering device that might be used
in the exhaust stream.

A combination of high mixing rate and long ignition delay to
approach premixed, homogenous conditions has been shown to reduce the
emission of particulates. However, periods of ignition delay that are
characteristic of today's diesel fuels are too short to achieve this,
using known mixing technology, and if longer mixing and ignition times
were permitted, the increased peak cylinder pressures that resulted
would cause serious structural problems. Thus, current approaches
optimize particulate emissions through programmed fuel injection and
system components that control the rate and intensity of combustion.
The kinetics of NO_x formation and the methods used to minimize the
amount of NO_x emitted are in direct opposition to the most effective
means for reducing the production of particulates in the combustion
chamber.

After 1984, achieving particulate emission levels below 0.5 g/mi
for cars of 2,000 to 4,000 pounds inertia weight and for trucks of more
than 2,500 pounds inertia weight, while simultaneously controlling
NO_x to 1.0 g/mi, may result in a cost disadvantage for diesel engines
relative to spark-ignition engines. The cost disadvantage will be
affected also by the likelihood that diesel fuel will not continue to
be cheaper than gasoline. What is more, the extra cost of a
particulate control device, as well as the costs of associated
materials and manufacturing methods, will decrease the economic
advantage of dieselization.

Test Methods

During the past five years, improved methods have been developed
for characterizing diesel engine emissions. The key method for
collecting and sampling diesel exhaust is a dilution tunnel system,
which was developed to simulate the exhaust dilution processes
occurring in the atmosphere. The system uses a particulate probe to
sample a diluted mixture of exhaust and air. During the engine
operating cycle, particulates are collected on a filter substrate and
undergo physical, chemical, and biological characterization. Current
particulate standards are concerned with only the total mass of the
particulates collected. However, the laboratory system makes it
possible to characterize the various chemical fractions contained in
the total mass sample. The material that can be extracted from the
particles with a conventional solvent is called the soluble organic
fraction. This gross fraction is important because it contains
biologically active components that can be further characterized
chemically.

In addition to determining the particulates, a separate probe, heated to 190°C (375°F), is used to extract gaseous emissions. Hydrocarbons are continuously monitored, but NO_x and carbon monoxide are determined later from a bag sample collected during the test cycle. Because the heated gaseous sample is taken upstream of the particulate sampling filter, a portion of the hydrocarbon emissions are accounted for once as gaseous hydrocarbon emissions and again as condensate on the particulate filter. This procedure was originally justified by EPA on the grounds that the particulate measurement was related to potential respiratory health effects and the hydrocarbon measurement was primarily related to smog effects. Thus, by double counting, the maximum potential effect was assessed for each. Hydrocarbons adsorbed on the particulate will appear in one set of effects or the other. They should therefore be counted only once.

Current EPA certification procedures specify that a maximum temperature of 52°C (125°F) should not be exceeded in the dilution tunnel sample zone. With different sized cars operating at various emission levels, a simple maximum temperature limitation does not provide mixture temperature control throughout the complete Federal Test Procedure driving cycle. Integrating the mixture temperature throughout the complete cycle, holding to a set limit (such as 30° $\pm$ 5°C) for the average mixture temperature, would provide more accurate results. In addition, although there would be excursions above and below a mean temperature, different sized cars would follow a similar mixture temperature profile throughout the Federal Test Procedure cycle. This should provide more comparable data for the various vehicles.

Because the dilution ratio is important for determining hydrocarbon adsorption and nitrogen dioxide levels (and in turn the chemical and biological character of the particulate sample), some temperature specification, such as the one suggested above, would be better for providing comparable data. Cycle temperature variations below 140°F do not appear to be significant relative to particulate mass measurements (Plee and McDonald, 1980).

Sulfates

In humid atmospheres, sulfates can hydrate to form sulfuric acid and acid sulfate mists, which can cause human health and environmental hazards, as well as destroy materials. Hence, sulfate emissions from diesel engines are of concern. Sulfates appear in diesel particulates as either acid "droplets" or more complex forms. There may be some free sulfur, but most of the sulfur will combine with carbon, hydrogen, oxygen, or nitrogen to form compounds such as hydrogen sulfide, alkyl sulfates, hetrocyclic compounds, sulfur dioxide, or sulfur trioxide. The relative amounts of individual species will depend on such factors as the amount of sulfur in the fuel, the mode of engine operation, the combustion temperature, the exhaust temperature, the dilution factor, and the overall particulate level. In the absence of a catalyst in diesel operation, the surface area of the diesel particulate, relative humidity, nitric oxide and nitrogen dioxide level, the air-fuel ratio, and the number of diesel particles may play an important part in sulfate formation (Khatri et al., 1979).

Visible Smoke

Visible smoke is an indication of a high exhaust particulate level. The smoke emissions of lightweight diesels, at conditions of full-load acceleration, are not federally regulated because the Federal Test Procedure is a relatively low-speed light-load cycle, with engine speeds ranging from 1,300 to 1,800 rpm at 10 to 25 percent of full load. Diesel engines generate their highest particulate concentrations and visible smoke under full-throttle operation and during cold-starting conditions. Diesel engines in heavy-duty vehicles are regulated for "worst case" visible smoke.

Manufacturers set their own product specifications for nonregulated conditions. Generally, a Robert Bosch smoke number of 3 is the maximum level permitted for full-load conditions. This number correlates to an 8 percent opacity when the Green and Wallace (1980) correlation is employed. Smoke values above 4 percent opacity are visible to the human eye.

Specifying the properties of diesel fuels can reduce the visible smoke from diesels. It is clear that worst-case (full-load acceleration conditions, cold start, and full-load lug) smoke limits will need to be developed to control visible smoke. Smoke control is likely to be necessary in order to gain widespread public acceptance of diesel cars.

Hydrocarbons

Atmospheric modeling studies performed so far show that methane does not contribute to the chemistry of smog formation. Because methane is a nonreactive hydrocarbon, any methane emissions could be excluded from all hydrocarbon pollution measurements for both diesel and gasoline engines. This approach would provide a larger adjustment for emissions from gasoline vehicles because new gasoline engines equipped with catalytic converters emit almost no reactive hydrocarbons. The hydrocarbon emissions from diesel vehicles are inherently low. Using the proper measurement technique (a particulate filter to divide the particulate and gaseous hydrocarbon phases), diesels have met the 1980 federal 0.41 g/mi standard (including methane) and California's 0.39 g/mi (non-methane) standard. Although gaseous hydrocarbons in diesel exhaust are low, the compounds emitted are highly reactive photochemically. Diesel fuel is less volatile than gasoline and its evaporative emission is extremely low--even without an emissions control system. Because of this, diesel vehicles could be given a credit in the federal hydrocarbon emission standard, as they are now in the California standard.

Carbon Monoxide

Carbon monoxide emissions from diesel engines also are inherently low. Consequently, even without emission control systems, diesels should always be able to meet the federal 3.4 g/mi carbon monoxide standard. In addition, carbon monoxide emissions from diesels do not increase significantly with the age of the engine, as is the case for catalyst-equipped gasoline engines.

Other Gaseous Species

Diesel engines emit more aldehydes, nitrogen dioxide, sulfur dioxide, and odors and irritants than gasoline engines. Of these four types of pollutants, nitrogen dioxide and the aldehydes are direct irritants. Nitrogen dioxide and other nitrogen compounds can react with the hydrocarbons to form biologically active nitro-organic compounds (Pitts et al., 1979). Better measurement methods, along with a better data base, are needed to characterize the tailpipe nitrogen dioxide levels. The odorous compounds consist of a variety of oxygenates and unburned fuel species. Both odor and irritants are increased by cold starting effects. Aldehydes, which are still unregulated, have high photochemical reactivities and are largely unaccounted for by current hydrocarbon emission assessments.

Noise

Diesels are typically noisier than gasoline engines, especially while idling and in low load. At such times they tend to clatter. The frequency range of the diesel's rackety sound seems to annoy more people than the pitch and cadence of properly operating gasoline engines. Available technology can be applied effectively to the engine compartment to make the din of diesels while idling and when passing by nearly as low as those of well-maintained gasoline engines. Efforts to reduce the characteristic sounds of diesels suggest that diesel noise is not a major long-term problem, although to suppress the noise levels may lead to higher costs or lower performance.

Fuel Effects

Data developed so far indicate that there is a relationship between the composition of diesel fuel and particulate and hydrocarbon emissions (Burley and Rosebrock, 1979). For instance, increasing the aromatic content and the 90 percent distillation temperature of the fuel will increase particulate and odor emissions. Using a minimum cetane number* and seasonally adjusted cloud point** will reduce

---

*A rating of the ability of diesel fuel to ignite quickly after it is injected into an engine cylinder; a measure of flash point.
**The temperature at which diesel fuel tends to become murky.

cold-start hydrocarbon emissions, noise, odor, and fuel system wax separation.

The feasibility of producing an improved diesel fuel needs to be studied to determine the trade-off between cost and availability of such a fuel and the potential engine and emission advantages. Data indicate that the following general limits may be desirable for improving the properties of diesel fuel for passenger cars, trucks, and buses:

Cetane Number	> 48
Aromatics	< 20 percent
90 percent Distillation Temperature	< 316°C (600°F)
Sulfur	< 0.25 percent by mass
Cloud Point Temperature	Seasonally adjusted

Current modifications to diesel fuel include control of fuel properties, fuel additives, and such nonconventional fuels and fueling configurations as emulsions, alcohol blends, fumigation, and synthetics.

Unfortunately, although improved specifications for an automotive diesel fuel would result in reducing particulate, sulfur dioxide, sulfate, and hydrocarbon emissions levels, as well as improving cold-starting characteristics, the more restrictive specifications might limit the available diesel fuel stock and increase refining costs and energy requirements. Furthermore, doing so might result in an indirect limitation on other distillate fuel production.

LUBRICATION

In addition to water contamination, diesel engines also have problems with lubricating oil contamination. The lubricating oil becomes contaminated when the products of combustion and the oil come into contact on the cylinder walls. Particulates appear to adsorb the additives that are used in lubricating oils to protect the surfaces subject to wear. The additives are substantially depleted when the particulate carbon content in the oil is at 2 to 3 percent by mass level. This problem is currently being prevented by shortening the intervals between oil changes and increasing the capacity of the oil sump. As it is, diesel engine manufacturers have suggested oil changes at intervals of every 3,500 miles—though one manufacturer now recommends a 7,000 mile interval. Manufacturers are attempting to correct the lubricating oil problems so that oil can be changed at intervals of at least 5,000 miles, with a goal of a 7,500 miles. Work also is under way to develop better oils.

ENGINE MODIFICATIONS

The combustion chamber geometry and the fuel injection system determine the combustion system of a diesel engine. Both must be optimized simultaneously. Table 2.3 shows how combustion chamber

TABLE 2.3 Effects of Combustion Chamber Design and Fuel Injection
Variables on Emissions.

| FACTORS* | GASEOUS EMISSIONS | | | PARTICULATES | |
	HC	NO$_x$	CO	UNBURNED HYDROCARBONS	SMOKE
Injection Timing	+++	+++	+	+++	++
Injection Rate	++	++	+	++	++
Spray Cone Angle	+	+	+	+	+
Secondary Injection	+++	+	+	+++	+
Combustion Chamber Geometry	+++	+++	+	+	+++
Injector Location	++	++	+	+	+

+++ Strongly Dependent

 ++ Moderately Dependent

  + Slightly Dependent

*All factors have been
optimized as much as
possible

design and fuel injection factors influence diesel emissions. (In the
table, particulates are broken down into hydrocarbons and smoke,
because hydrocarbons contribute to the soluble organic fraction and
smoke is an indicator of the particulate fraction.)

Emissions data demonstrate the importance of injection timing in
minimizing hydrocarbons, NO$_x$, and particulates. Engine emissions and
fuel consumption cannot be simultaneously minimized at a single timing
point for the diesel. The pump/line/nozzle injection system, modified
for improved timing and injection control as a function of load and
speed, will continue to be used in light-duty vehicles through 1984.
Other control functions on current light-duty vehicle engines are an
automatic cold-start timing advance, a temperature-controlled idle stop
(for fuel quantity change), and a temperature-controlled quantity of
starting fuel. Turbocharged engines also use aneroid devices that
sense the reduced engine boost pressure during acceleration and lower
the full-load fuel delivery rate until the boost pressure reaches a
preset level.

Electronically controlled pumps, which are designed to reduce
hydrocarbon, NO$_x$, and particulate emissions, are currently in

research and development and should be available in 1985 model cars. With recent developments in microprocessors, electronic control of fuel injection is being pursued and is compatible with the 1984 all-altitude emission requirement. In addition to the technical advantages, the electronically controlled fuel pump requires a smaller capital investment than the advanced technology for fully controlled mechanical systems. Therefore, it has an economic advantage.

With this system, signals proportional to fuel rate and piston position are measured by sensors and are processed by an electronic control system to determine the optimum fuel rate and timing (Gliken et al., 1979). The maximum fuel delivery curve as a function of speed, and the timing map as a function of speed and load (or derivatives), are stored in the programmed memory of the microprocessor. This makes it possible to closely match the fuel injection with operating engine demands or previous operating conditions.

The technology for such a system is here. A number of prototype cars are currently running with electronic fuel injection control systems. However, further development needs to be carried out on sensors and actuators to make the technology available for production by 1985. Thus, the all-altitude system needed to meet the 1984 federal standard would be an interim mechanical system that would be used one year only. Under the circumstances, it may be reasonable to delay implementing an all altitude requirement until the 1985 model year, when electronic fuel control systems should be available.

Turbocharging

Blowing or turbocharging diesel engines is an effective way to increase the power output from an engine of given displacement. This technology is attractive because it enables manufacturers to add additional power levels to a basic displacement engine. If turbochargers are properly designed and developed, in conjunction with vehicle design (i.e., engine installation and rear axle ratio), improvements in emissions can also be achieved.

Turbocharger technology is certain to be more widely used in automobile diesel engines in the early 1980's. One of the limiting factors in current turbochargers is their poor low-speed efficiencies. Improving low-speed efficiency, which can improve the acceleration performance of turbocharged indirect-injection engines, is an important area for research and development. Turbocharging is also important for direct-injection engines because the smoke-limited load/speed range of naturally aspirated direct-injection systems is usually less than that of indirect-injection engines.

Direct Injection

Direct-injection diesel engines offer about a 15 percent fuel economy advantage over current indirect-injection engines. Thus, light-duty direct-injection engine development is being pursued vigorously by a number of automobile manufacturers and other organizations. Despite the fuel economy advantage of the direct-injection diesels, all diesel-powered cars now being sold use indirect-injection engines. Current direct-injection engines have lower speed and power output capabilities, higher hydrocarbon, NO_x, noise, and odor emissions, and higher peak cylinder pressures and weights than indirect-injection engines.

Efforts are under way to increase direct-injection engine output by increasing the combustion rate. Reductions in hydrocarbon, odor, and noise emissions are being sought by increasing the turbulence level, retarding injection timing, minimizing the injection-nozzle sac volume, and increasing the compression ratio (Cartellieri, 1980). It appears that turbocharging may be needed to increase the smoke-limited load/speed range, which is generally higher in indirect-injection engines.

To overcome the current limitations, the development of direct-injection diesel engines needs to be accelerated to attain the advantages projected in Table 2.4.

Exhaust Gas Recirculation

Exhaust gas recirculation (EGR) is the principal engine technology being used to reduce NO_x emissions. EGR reduces combustion temperatures and thereby NO_x emissions. However, EGR, if improperly applied, can increase the amount of particulates. EGR systems fall into two general categories--constant and modulated. In the long run, modulated electronic EGR systems appear to be preferable, but simpler mechanical systems will be used during the early 1980's. These will enable the 1982 to 1984 emissions standards (0.41, 3.4, 1.0, and 0.6 g/mi for hydrocarbon, carbon monoxide, NO_x, and particulates, respectively) to be met, with some waivers to 1.5 g/mi for NO_x, at either no cost or a small cost for incremental controls, in comparison with 1981 model year vehicle standards. Modulated EGR systems range from simple on-off systems (with relatively constant EGR for low to moderate loads and no EGR for heavy loads), to sophisticated closed-loop electronically controlled devices that adjust the EGR rate to optimize NO_x, HC, and particulate emissions, as well as fuel economy.

To minimize all diesel emissions, engine timing and other design variables need to be reoptimized to take advantage of EGR. Because it reduces the oxygen-fuel ratio, EGR can increase CO and particulate emissions. Air flow, engine speed, and fuel pump load sensors are used to accurately determine the air-fuel ratio under all operating conditions. An accurate CO_2 sensor is another approach for measuring a single variable to control the air-fuel ratio and EGR--though no reliable sensor is now available.

TABLE 2.4 Evaluation of Direct Injection in Relation to
          Indirect Injection.

SUMMARY

DI IN RELATION TO IDI

	Better	Equal	Inferior
Fuel Economy	15-20 <		
Heat Reduction To Coolant	-30 <		
Starting	_____		
Suitable For Turbocharging	_____		
Speed Range		_____	
BMEP Potential		_____	
Smoke Full Load			_____
Smoke Part Load	_____		
Low NO_x/HC Potential (USA Only)			_____
Particulates	_____		
Noise (Combustion)			
- Full Load			_____
- Idle	_____		
Durability		_____	
Production Cost		_____	

(Source: Cartellieri, 1980).

EXHAUST AFTER-TREATMENT

The development of diesel engines with low particulate emissions is now constrained by fuel quality, engine design, government NO_x standards, materials, and the driving patterns of motorists. Important nonregulated air pollutants from diesel engines are aldehydes, NO_2, SO_2, the particulate soluble organic fraction, and odor. Aldehydes and NO_2, which diesels emit in greater quantity than gasoline engines, particularly in cold start-up conditions, are photochemically active and act as direct irritants and odorants. Diesel emissions of SO_2 and sulfates are proportional to the sulfur levels in the fuels. Some proposed controls for exhaust components decrease sulfur dioxide emissions and increase sulfate emissions.

Control technologies using catalysts may prove effective for reducing the soluble organic fraction of particulate, HC, aldehyde, and SO_2, though increased levels of sulfates and NO_2 are likely to result. Indeed, HC, NO_x, and particulate deterioration factors could worsen with emission controls.

Because diesels inherently emit small amounts of CO, they should easily meet the 3.4 g/mi CO standard for light-duty vehicles. But the reduction of particulate emissions from diesels is dependent on the development of reliable, durable, and marketable regenerative trap-oxidizers or some other control device, as well as advances in engine design and properties of diesel fuels. What is more, vehicle weight will determine in large measure the minimum level of particulate, which could approach 0.1 g/mi for the smallest diesel cars.

The development of after-treatment technology for limiting particulate emissions from diesel engines must take into account the low and variable exhaust temperatures in light-duty vehicles. Typically, exhaust temperatures range from 150°C to 300°C over the entire cycle of the Federal Test Procedure. Exhaust temperature is approximately linear with load. It increases with the engine speed and is highly transient while the diesel is put through the Federal Test Procedure (Toyota, 1980). This is because the diesel has a variable air-fuel ratio and the test cycle requires substantial low-speed, part-load operation. By contrast, the gasoline engine, which operates essentially with a constant air-fuel ratio, produces exhaust temperatures that range from 600°C to 700°C during the test cycle.

Three basic approaches to exhaust after-treatment are under investigation:

- Reactors/thermal in-stream oxidation;
- Catalysts; and
- Traps, catalyzed traps, and trap-oxidizers.

Of these approaches, catalyzed or uncatalyzed trap systems appear to be the most promising. They will be designed to regenerate automatically every 50 to 100 miles. A trap is usually only used to remove particles from the exhaust stream; it also can be used to concentrate particulate matter. The three principal types of materials under consideration are metal meshes, ceramic monoliths, and ceramic foams. To a lesser degree, paper materials and other fibers are also under investigation. Figure 2.3 shows two design approaches to after-treatment filters.

An overall engine trap system for post-1985 application is illustrated in Figure 2.4. This system uses an intake throttle regenerative system with electronic fuel injection and two traps in series. At the present stage of development, it is not possible to provide a detailed description of an optimum system for controlling diesel exhaust.

Before introduction into the marketplace, diesel emission control technologies must be evaluated for:

- Effects on NO_x, hydrocarbons, carbon monoxide, and particulates;
- Effects on unregulated emissions;
- Effects on fuel economy;
- Effects on vehicle acceleration and driveability;
- Effects on engine durability;
- Need for active control of concept and the degree of sensitivity of control;
- Complexity;
- Degree of maintenance required;
- Relative cost; and
- Ease of integration within the engine.

DIESEL FUELS

Commercial diesel fuels are mixtures of hydrocarbons derived from crude oil. The properties of diesel fuel depend on the types of crude oils used as raw material, the refinery process, and the properties and mixtures of the refinery stocks from which the fuel is blended. Variations in such factors affect the ignition quality, volatility, calorific value, hydrocarbon composition (e.g., paraffins, naphthenes, olefins, or aromatics), sulfur content, and other properties.

The specification of a minimum quality automotive diesel fuel would enable engine designers to optimize diesel engines and would ensure that the performance of existing engines would not degrade. Important specifications to be considered are a minimum cetane index, a maximum sulfur level, a maximum aromatic content, a maximum 90 percent boiling point, and a seasonally adjusted cloud point. As the future quality of crude feedstocks falls, fuel with lower cetane numbers, higher aromatic and sulfur content, and higher end-points may be expected. The establishment of tighter specifications for automotive diesel fuel will result in restricting refining flexibility to manufacture a range of products from low quality crudes. Thus, to optimize diesel fuel availability, efforts need to be made to improve the fuel tolerance of the diesel engine.

There is wide variation in the properties of diesel fuels marketed in the United States. Improved analysis of the extent and importance of these variations is needed.

Figure 2.3 Two Types of Exhaust After–Treatment Filters

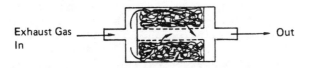

Metal Mesh Filter

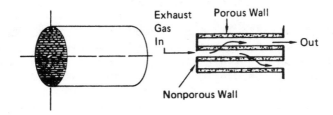

Exhaust Gas
In

Out

Ceramic Filter

Exhaust
Gas
In

Porous Wall

Out

Nonporous Wall

Figure 2.4 Emission Control System for Diesel-Powered Cars

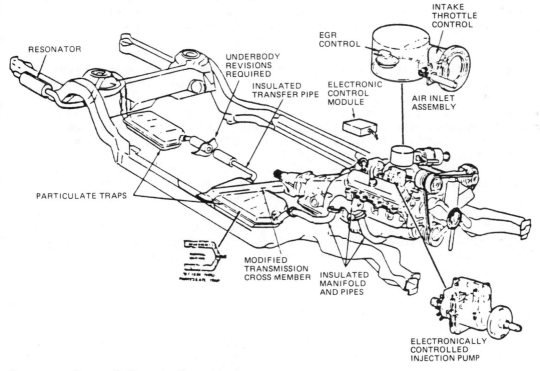

Source: General Motors Corporation

The oil industry now produces two grades of automotive diesel fuel--Nos. 1 and 2, established under ASTM 975 specification. No. 1 diesel fuel is produced primarily for circumstances in which No. 2 causes cold weather handling and engine starting problems. No. 1 diesel fuel is also used in some city buses to reduce smoke emissions. The price of No. 1 fuel is currently higher than that of No. 2. The availability of No. 1 fuel for passenger cars is expected to decrease in the future because of competing demands for its use as blending stock and jet aircraft fuel. Even now, No. 2 diesel fuel is most commonly used in trucks and passenger cars and in the future may be the only diesel fuel available in large quantities.

Demand and Supply

If sales of diesel cars and small trucks reache 25 percent of total annual sales of light-duty vehicles by 1990, the demand for diesel fuel in that year will be about 440,000 barrels per day. This quantity of fuel represents only 13 percent of the 1979 consumption of all middle-distillate fuels.

A survey of the petroleum industry indicates that there will be no major problems supplying increased demands for current ASTM 975 specification No. 2 diesel fuel, even if sales of diesel light-duty vehicles represent 50 percent of light-duty vehicle sales by 1990. The supply of diesel fuel will be achieved by reducing the conversion of distillates to gasoline and by adding a limited amount of refinery facilities. Additional further supplies could be made available through the conservation of heating oils.

Assuming sharply increased dieselization of intermediate-sized trucks and continued high use of large diesel trucks, highway diesel fuel consumption is expected to double by 1990 from the 850,000 barrels per day consumed in 1979. Thus, the estimated demand for diesel fuel for light-duty vehicles during the next ten years is relatively small in the context of total distillate demand, including the requirements of highway trucks. Table 2.5 shows the 1978 demand for petroleum products and the projected demand for 1990.

Published studies indicate that a modern oil refinery will consume less energy per barrel of capacity as the proportion and volume of diesel fuel production are increased. Most of the lower energy consumption in refineries occurs because of reduced catalytic cracking operations. As a typical modern refinery shifts from maximum gasoline production to maximum distillate output, the energy in the fuel, as a percent of energy input as raw material, will increase by 1 percent from about 92 to 93 percent of input energy (Lawrence et al., 1980).

Fuel Price

The cost of crude oil is the major factor determining the cost of producing diesel fuel. Typically, the crude oil raw material cost may represent 85 to 90 percent of the cost of producing diesel fuel.

TABLE 2.5 Domestic Petroleum Product Demand, TBD.

	Actual 1978	Projected 1990
Motor Gasoline	7,412	6,124
Jet Fuel		
Naphtha Type	199	124
Kerosene Type	858	1,203
Distillates		
Automotive Diesel[1]	988	1,988
Other[2]	2,629	2,287
Residual Fuel Oil	3,023	2,344
Other[3]	3,748	4,827
TOTAL DEMAND	18,857	18,897

[1] Includes off highway use.

[2] Includes heating oils for households, industry, electric
     utilities, railroads, vessels, military, and miscellaneous.

[3] Includes aviation gasoline, naphthas, liquefied gases,
     petrochemical feed stocks, lubricants, waxes, coke,
     asphalts, road oils, still gases for fuel, and miscellaneous.

(Source: National Petroleum Council [preliminary unpublished data])

Although a minor part of the total, the other significant element is
the refining cost, which is inherently lower for diesel fuels as
compared with gasoline. Because of the unpredictability of future
crude oil availability and pricing policies determined by foreign
producers, the committee has not attempted to forecast the impact of
raw material costs on diesel fuel prices.

The relative prices of diesel fuel and gasoline are an important
consideration in this study. Although the price difference has
narrowed in the past five years, diesel fuel in the previous ten years
(1967 to 1976) sold at a price that was about 15 percent less than the
price of regular gasoline (American Petroleum Institute, 1981).

Confronted by declining gasoline demand and increasing distillate demand, the oil companies predict a narrowing of the differential between gasoline and distillate prices. Assuming free market conditions, the refiners forecast that some time in the 1980's the price of diesel fuel will equal the price of unleaded gasoline. During the same period, kerosene jet fuel prices are expected to rise above the price of gasoline. To maintain refinery profit margins, the refiners will shift a larger proportion of their total costs onto the increasing middle-distillate production. This shift will have the effect of lessening the current operating cost advantage of diesel-powered light-duty and heavy-duty vehicles relative to gasoline-powered vehicles.

REGULATORY ISSUES

The EPA test procedure is adequate for determining the total mass and soluble organic fraction of diesel particulate emissions--though the adequacy of the test method for identifying the individual hydrocarbon species and biological activity of the particulate soluble organic fraction remains to be established. Current emission test procedures result in "double counting" the heavy portion of hydrocarbon emissions from diesel engines. This could be rectified by changing the test procedure to measure the heavy hydrocarbons as particulates only.

Significant differences exist in the characteristics of hydrocarbon emissions from diesel and gasoline engines. Such differences need to be reflected in establishing appropriate hydrocarbon emission regulations. A nonmethane hydrocarbon emission standard for both engine types and credit for lower evaporative hydrocarbon emissions from diesel vehicles ought to be considered. Current test methods do not correctly account for aldehydes as hydrocarbon emissions.

Meeting the 1982 to 1984 federal light-duty vehicle emission standards is technologically feasible, assuming that some NO_x waivers are granted for larger diesel cars in 1983 and 1984. Some 1981 diesels already meet the EPA particulate standards for passenger cars and small trucks in the model years 1982 to 1984 without the use of particulate control devices. Others will require improved exhaust gas recirculation systems. The cost of this is estimated to be from nothing at all to $30 for each diesel light-duty vehicle.

Meeting the 1985 NO_x and particulate emission standards of 1.0 and 0.2 g/mi, respectively, for light-duty diesels is technologically feasible for only the smallest vehicles of 2,000 pounds or less. Larger cars are not likely to meet the particulate standard without an effective emission control device. An NO_x standard of 0.4 g/mi will probably not be met except for the smallest diesel vehicles. Control of particulate at the 0.2 g/mi level is likely to require the use of exhaust after-treatment devices. One of the most promising is the regenerative trap-oxidizer--though it has not yet been proven in field durability tests of 50,000 miles, nor has it met many of the requirements considered essential to commercialization for passenger cars in use.*

The experience of developing gasoline catalyst systems for 1975 model passenger cars suggests about a five-year lag between the design and demonstration of such a device and its production and commercialization. The lesson here is that development takes longer than anyone expects. The inventor tends to underestimate the testing time. For some types of products--and emission control devices for something as universal as the automobile is an example--a realistic trial period using a test fleet is essential to reveal problems that cannot be anticipated in development. The mechanical and economic performance of a complex and sensitive system like the diesel engine cannot be safely predicted from laboratory demonstration or theoretical analysis, but only from prolonged experience in actual use.

Opportunities for controlling particulates in diesel heavy-duty vehicles and limiting NO_x emissions from gasoline-powered vehicles need to be considered by the EPA in developing optimum emissions control strategies for light-duty diesel cars. Delaying the 1984 high-altitude emission requirement for diesel-powered light-duty vehicles until 1985 would eliminate the need for developing an interim fuel injection system that would be good for only one model year.

---

*At present there are at least four different approaches under development to control diesel particulate emissions.

REFERENCES

Amann, C. A., D. L. Stivender, S. L. Plee, and J. S. MacDonald (1980). Some Rudiments of Diesel Particulate Emissions. SAE Technical Paper 800251.

American Petroleum Council (1981). Unpublished data.

American Petroleum Institute (1980). Basic Petroleum Data Book. API: Washington, D.C.

Burley, H. A., and T. L. Rosebrock (1979). Automotive Diesel Engines – Fuel Composition vs Particulates. SAE Technical Paper 790923.

Carpenter, K., and J. H. Johnson (1979). Analysis of the Physical Characteristics of Diesel Particulate Matter using Transmission Electron Microscope Techniques. SAE Technical Paper 796815.

Cartellieri, W. (1980). Direct Injection For Light Duty Diesel Engines. Manuscript. La Societe des Ingeiniers de l'Automobile.

Green, G. L., and D. Wallace (1980). Correlation Studies of an In-Line, Full Flow Opacimeter. SAE Technical Paper 801373.

Gliken, P. E., D. F. Mowbray, and P. Howes (1979). Some Developments on Fuel Injector Equipment for Diesel Engine Powered Cars. I. Mech. E.

Khatri, N. J., J. H. Johnson, and D. G. Leddy (1978). The Characterization of the Hydrocarbon and Sulfate Fractions of Diesel Particulate Matter. SAE Technical Paper 780111.

Lawrence, D. K., D. A. Plantz, B.D. Keller, and T. O. Wagner (1980). Automotive Fuels--Refinery Energy and Economics. SAE Technical Paper 800225.

Murrell, J. D. J. A. Foster, and D. M. Bristor (1980). Passenger Car and Light Truck Fuel Economy Trends through 1980. SAE Technical Paper 800853.

Nissan Motor Company (1980). Briefing to the Diesel Impacts Study Committee Technology Pane.

Pitts, Jr., J. N., K. A. Van Cauwenberghe, D. Grosjean, J. P. Schmid, D. R. Fritz, W. L. Belser, Jr., G. B. Knudson, and P. M. Hynds (1979). Atmospheric Reactions of Polycyclic Aromatic Hydrocarbons: Facile Formation of Mutagenic Nitro Derivatives. Science 202:515.

Plee, S. J., and J. S. McDonald (1980). Some Mechanisms Affecting the Mass of Diesel Exhaust Particulates Collected Following a Dilution Process. SAE Technical Paper 800186.

Toyota Motor Company (1980). Briefing to the Diesel Impacts Study Committee Technology Panel.

U.S. Environmental Protection Agency (1980). 1980 Gas Mileage Guide Second edition.

U.S. Environmental Protection Agency (1981). 1981 Gas Mileage Guide.

Vuk, C. T., M. A. Jones, and J. H. Johnson (1976). The Measurement and Analysis of the Physical Character of Diesel Particulate Emissions. SAE Technical Paper 760131.

3 ENVIRONMENTAL EFFECTS

The environmental effects of increasing the use of diesel engines in passenger cars and light trucks are analyzed in this chapter by characterizing the emissions, describing the processes and changes that occur when the emissions are mixed in the atmosphere, and evaluating the known and potential consequences of the transformed and primary emissions on visibility, climate, ecology, and human health.

Light diesel engine exhaust contains both carbon particles and gaseous substances. They are characterized here in terms of their physical and chemical properties at the tailpipe and compared quantitatively to those of conventional gasoline engine emissions. The several components of diesel exhaust undergo various processes once they are released into the atmosphere. Primary pollutants are dispersed and physically and chemically transformed into secondary substances. The reaction products in turn are transported in the atmosphere for various distances and ultimately affect the quality of the air, visibility, and possibly even the climate in certain urban areas.

While a completely reliable tracer or signature specific to diesel emissions has not yet been identified, simulations and models for various measures of exhaust dispersion have been developed. Air quality assessments have been undertaken from the perspectives of many scientific disciplines, though research into the environmental effects of emissions produced specifically by light diesel engines is only in its beginnning stages. Little data are available specific to diesel-generated pollutants in terms of air quality or overall ecological consequences. Even so, sufficient information exists to describe the chemical and physical character of diesel emissions, to evaluate projections of their reactions with other substances in the atmosphere, and to suggest productive directions for future research.

The factors considered in this analysis of environmental effects are outlined in Figure 3.1. Emissions are first characterized in terms of parameters relevant to air quality. Second, the fates of these emissions are examined in terms of the ambient loadings of primary exhaust pollutants and their potential chemical transformations in the atmosphere. Third, the potential effects are discussed in terms of visibility, ecology, and health.

Figure 3.1 Relating Environmental Quality to Diesel Engine Emissions

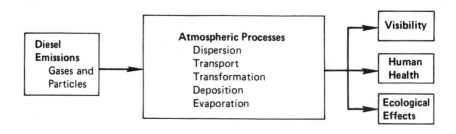

CHARACTERIZATION OF EMISSIONS

Diesels, like gasoline engines, emit both particulate and gaseous pollutants as products of combustion. The important distinction between the two is their differing rates of emissions. Diesels emit from 30 to 100 times more particulate matter than comparable gasoline engines equipped with catalytic converters. Although the same federal standards currently regulate emission levels for both types of engines, the individual components of diesel hydrocarbon and nitrogen oxide emissions may have environmental consequences different from those of gasoline engines.

PARTICULATE EMISSIONS

Diesels emit particles at a much greater rate than gasoline engines, and the size distribution of these particles has important potential consequences for the environment as well as for public health. Most of the particulate matter emitted from diesel-powered light-duty vehicles consists of submicron carbonaceous agglomerates 0.06 to 0.7 μm in diameter. Emission rates for light diesels have been measured in the range of 0.2 to 0.8 g/mi (Springer, 1978; Pierson, 1978). This is far greater than the 0.02 g/mi emission rate for particles from comparable catalyst-equipped spark-ignition engines. The size distribution of diesel exhaust particles is important because transport of particles in the atmosphere and deposition in the human respiratory tract depend essentially on size (Lippman, 1976).

The size distribution of diesel exhaust particles is compared to those of typical urban atmospheric particles in Figure 3.2. In the atmosphere, particles occur in three modes: a nuclei mode (usually 0.005-0.1 μm in diameter), an accumulation mode (0.1-2.0 μm), and a coarse particle mode (2.0-50 μm) (Whitby, 1978). The nuclei mode is associated with nucleation of low vapor-pressure materials such as sulfuric acid, lead salts, or carbon. Particles in this mode tend to grow quickly by condensation and coagulation. The relative concentration in the nuclei mode depends on the proximity to the source of emissions. The accumulation mode is formed both by coagulation of nuclei mode particles and by nucleation and condensation of moderate vapor-pressure materials such as the products of photochemical smog. The mode may also be fed by heterogeneous gas-to-particle conversion processes. In the stable accumulation mode, further growth is greatly retarded and such removal processes as settling and deposition are slow. The coarse particle mode includes windblown dust and roadway debris that do not interact strongly with particles in either of the other two modes.

The volume-weighted size distribution of the particles emitted from light diesel engines is not strongly dependent on engine type and fuel characteristics. Most of the particles emitted from light diesels are in the accumulation mode and possess a mean diameter of about 0.2 μm. Only a few percent of diesel particles are in the nuclei mode. Coarse particles rarely exceed about 15 percent of the total, and the amount of coarse particles emitted appears to depend essentially on the engine operating cycle, with larger emissions occurring during transient conditions (Hare and Baines, 1979).

Diesel-emitted particles have high specific surface areas of 30 to 100 m^2/g, which are similar to those typical of activated carbon and much larger than those of ambient aerosols. They are capable of adsorbing relatively large quantities of organic material; the solvent extractable fraction is typically 5 to 40 percent, but it may be as high as 90 percent (Williams and Begeman, 1979; Funkenbusch et al., 1979). The elemental composition of the extractable material averages 81 percent carbon, 12 percent hydrogen, 6.8 percent oxygen, 0.4 percent nitrogen, and 0.4 percent sulfur. On the average, 55 percent of the extractable material consists of aliphatic, olefinic, and alicyclic hydrocarbons (with a carbon number maximum at 20 to 22) from unburned fuel and lubricating oil (Black and High, 1979; Mayer et al., 1980). About 10 percent consists of polycyclic aromatic hydrocarbons; the unsubstituted hydrocarbon is accompanied by several of its alkyl homologues. Benzo[a]pyrene, a carcinogenic polycyclic aromatic hydrocarbon, has a mass concentration of 0.3 to 0.6 $μg/m^3$ and an emission rate of 2 to 5 μg/mi. On average, 25 percent consists of oxygenated compounds such as polycyclic aromatic quinones, aldehydes, fluorenones, and naphthalic-anhydride; in most cases the unsubstituted compound is accompanied by alkyl substituted homolog with up to six

Figure 3.2   Comparison of Typical Atmospheric and Diesel Exhaust
Aerosol Size Distributions. Upper figure represents the
size distribution of a typical atmospheric aerosol, and
lower plot shows size distribution of typical diesel
exhaust particulates.

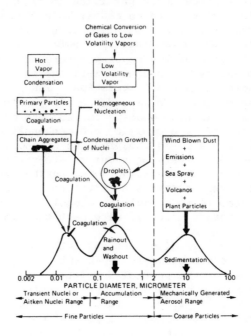

Source: Whitby, 1978

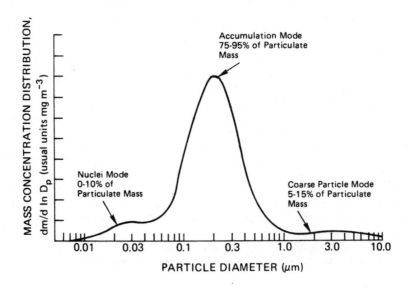

alkyl carbon atoms. About 10 percent are acids, and about 0.5 percent are bases.

Optical Properties

Diesel particles are light absorbers, with a specific extinction coefficient three times larger than that of typical urban areosols. Lipkea et al. (1978) have determined the optical extinction coefficients, from smoke opacity and mass concentration measurements, of 8.2 m^2/g for raw diesel exhaust particles. This extinction coefficient is the sum of adsorption and scattering components.*

Absorption cross-sections over actual roadways (Gorse, 1980) are 6.7 m^2/g based on total mass. Laboratory measurements by Sherrer et al. (1980) and Japar and Szkarlat (1980) give, respectively, cross-sections of 9.2 m^2/g and 8.7 m^2/g based on nonvolatile mass. These results suggest that light absorption by diesel particles is influenced mainly by the carbon content and that adsorbed species have little effect on absorption.

If it is assumed that 75 percent of the particles characterized by Lipeka et al. (1978) are nonvolatile and that absorption is 80 percent of extinction, an absorption cross-section based on nonvolatile mass may be calculated at 8.7 m^2/g. If the same assumption about the nonvolatile fraction is made for Gorse's (1980) data, a value of 8.4 m^2/g is obtained. Thus, the different measurements of light absorption cross-sections based on nonvolatile particle mass agree very well, with values ranging from 8.4 to 9.2 m^2/g. Studies of the light-scattering cross-section based on total particle mass are not in such good agreement; values ranging from 1.3 to 4 m^2/g were reported (Gorse, 1980; Pierson, 1978; Sherrer et al., 1980). Waggoner (1980) shows that light scattering depends more on size distribution than does light absorption. This may be the reason for the greater variation.

Diesel Exhaust Aerosol Dynamics

The physical and chemical properties of diesel particles in the atmosphere depend on processes taking place not only in the engine and exhaust system but also in the exhaust plume and the atmosphere. Measurements made in the exhaust plumes of diesel cars show that dilution ratios of several thousand to 1 are reached in less than 1 second. However, most laboratory studies of diesel aerosols produce dilution ratios between 5 and 20 to 1 (typically 13:1) and have transit times of 1 to 5 seconds. These diluted streams may then be passed through impactors and/or filters for sampling and analysis or to exposure chambers for animal studies. The total time the particles are situated in such a system may be many minutes.

---

*All results reported here are for light in the 500 to 550 nm wavelength range.

Diesel aerosols are highly dynamic nonlinear systems. Processes taking place in the relatively concentrated streams produced by laboratory dilution systems may lead to particle size distributions and extractable fractions different from those produced under roadway dilution conditions. These include coagulation, condensation/adsorption, evaporation/desorption, and chemical reactions—each described below.

Particle growth by coagulation results from interparticle collisions and thus depends strongly on dilution. Coagulation half-lives, which are directly proportional to dilution ratios, range from about 1 second in undiluted exhaust to about 10 seconds in a typical dilution tunnel to about 1,000 seconds over a roadway. Animal exposure chambers are often charged with diesel aerosols piped from dilution tunnels at low dilution ratios; the chambers have aerosol residence times of many minutes. Coagulation in these conditions may significantly change the size distribution of the particles and the results of such experiments should be interpreted with care.

The periods such particles hover over roadways, however, are short compared to coagulation half-lives, so that little change from the size distribution at the tailpipe would be expected. As coagulation continues, though, these aerosols ultimately become part of the atmospheric aerosol but at a much reduced rate.

Condensation/adsorption and evaporation/desorption are considerably more difficult to quantify, because neither the condensing species nor their concentrations are known. Kittelson and Dolan (1980) and Johnson et al. (1978) have suggested that the extractable fraction of diesel exhaust particles collected from dilution tunnels should depend on the dilution ratio. Data on the influence of dilution ratio on the organic extractable fraction are limited and somewhat contradictory. Frisch et al. (1979) and MacDonald et al. (1980) show a slight dependence of extractables on dilution ratio; Black and High (1979) and Bradow et al. (1979) show none. Thus, the question of the influence of dilution ratio on the extractable fraction has not yet been resolved, and the results of studies done at moderate dilution ratios should be interpreted with care. This problem is further complicated by considerations of the kinetics of gas-to-particle mass transfer.

Adsorption kinetics have not been measured for diesel particles. However, Natusch and Tomkins (1978), Rothenburg and Cheng (1980), and Fuller et al. (1978) have examined the kinetics of adsorption by fly ash. They report characteristic adsorption times ranging from a few seconds to many minutes. If diesel particles behave similarly, adsorption kinetics will influence the gas-to-particle transfer process under roadway conditions, where there is little time to reach equilibrium. It is under these conditions that the general public is likely to be most heavily exposed. Unfortunately, it is difficult to determine experimentally whether departures from equilibrium occur, because the act of filtration holds the particles in contact with the roadway or dilution tunnel gas phase constituents for many minutes. The partitioning of adsorbable materials between the gas and particle phases may have an important bearing on their environmental effects.

GASEOUS EMISSIONS

Although the gaseous hydrocarbons emitted by diesel engines (e.g., ethylene, propylene, and formaldehyde) are inherently low, they are by mass 80 to 90 percent photochemically reactive materials, including 30 to 35 percent carbonyl compounds that are potent precursors of smog. (In this discussion, nonreactive compounds include ethane, acetylene, propane, and benzene.) The vapor phase of diesel exhaust contains, on average, [2 ppm methane] 4 ppm acetylene, 20 ppm ethylene, 4 ppm propylene, 2 ppm 1,3-butadiene, 2 ppm 1-butene, 12 ppm formaldehyde, 4 ppm acrolein, and 1 ppm or less of at least 50 other volatile hydrocarbons and aldehydes. The emission rates of some of these compounds are: [methane, 13 mg/mi] ethylene, 43 mg/mi, propylene, 14 mg/mi, formaldehyde, 20 mg/mi, and acrolein, 10 mg/mi.

By contrast, new spark-ignition vehicles equipped with catalytic converters emit almost no reactive hydrocarbons in their exhausts. In terms of total emissions, however, catalyst deterioration and the growth in evaporative emissions over the lifetime of the vehicle might cause an increase in the amount of reactive material released. All the evaporative emissions from spark-ignition engines are reactive, while diesels have negligible evaporative emissions. Although the same federal emission standard of 0.41 g/mi applies to exhaust hydrocarbons for both diesel and spark-ignition light-duty vehicles, diesel engines emit more reactive hydrocarbons. In addition, the potential for vapor escaping from the fuel distribution system are significantly greater for gasoline than for diesel fuel.

The 1981 federal emission standard for nitrogen oxides (NO_x) is 1.0 g/mi for both diesel and spark-ignition light-duty vehicles. However, about 18 percent of the NO_x in diesel exhaust consists of the more hazardous NO_2, while spark-ignition engines emit less than 10 percent of the oxides as NO_2. The remainder in each case is nitric oxide (NO). Moreover, there is a question of deterioration in the NO_x emission control in both types of engine systems. For instance, the performance of diesel exhaust gas recirculation systems (EGR) may decline if valves are inadequately maintained. Even so, the assumption of no deterioration is reasonable. In spark-ignition systems, the EPA assumes, control system deterioration that tends to increase NO_x emissions overshadows the engine deterioration that tends to decrease NO_x.

Sulfur dioxide (SO_2) emissions from diesels are 5 to 10 times higher than those from gasoline engines because of the larger sulfur content of diesel fuel. However, carbon monoxide (CO) emissions from diesel engines are lower, near 1 g/mi, compared with 1 to 15 g/mi for catalyst-equipped spark-ignition automobiles.

AIR QUALITY EFFECTS OF PRIMARY EMISSIONS

Primary diesel emissions can affect air quality in several ways. Particulate emission levels can be estimated by comparing CO emission rates and atmospheric concentrations of CO. Visibility reductions can be calculated from projected ambient loadings, and atmospheric heating rates in urban areas can be estimated from diesel additions to airborne graphitic carbon. Reductions in CO levels with increased diesel use and increases in other gaseous emissions can be predicted.

Particulates

The effects of greater diesel use on air quality can be estimated by using CO as a tracer, or indicator, of particulate dilution rates. The CO emission rates of the existing light-duty vehicle fleet and atmospheric concentrations of CO can be used to calculate the dilution of light-duty vehicle emissions. The key assumptions are that CO dispersion can approximate particulate dispersion for vehicle emissions, the spatial distribution of all CO emissions will be the same as that of CO emissions from spark-ignition vehicles, the fraction attributed to light vehicles is known, and traffic patterns will be unaltered by growing diesel use.

Assuming these relations, particulate levels ($\mu g/m^3$) resulting from the substitution of diesel vehicles for spark-ignition vehicles are calculated from the expression

$$\text{Particulate} = f \frac{Q'_{Part}}{Q_{CO}} (CO)_{amb, LDV}, \tag{1}$$

where Q'_{Part} is the emission rate (mg/mi) of particulate matter from a new diesel relative to those of a standard spark-ignition vehicle, Q_{CO} is the CO emission rate (mg/mi) averaged over the percentage of vehicle miles driven for each model year of light vehicles within each model year in a specific urban area, $[CO]_{amb}$ is the ambient concentration of CO ($\mu g/m^3$) from light vehicles, and f is the fraction of diesels in the whole light-duty vehicle (LDV) fleet. Equation (1) applies to primary emissions, but not to such secondary conversion products as the sulfate, nitrate, and organic components of photochemical aerosols (Grosjean and Friedlander, 1975).

Urban Ambient Aerosol Mass Loadings

Increases in benzo[a]pyrene levels can be obtained by substituting the benzo[a]pyrene emission rate for diesels relative to that of spark-ignition light-duty vehicles for Q'_{Part} in equation (1).

Calculations for air quality impacts were carried out for four urban areas with disparate settings--Los Angeles, Phoenix, St. Louis, and New York City, or, more specifically, Manhattan. Annual hourly average CO concentrations were obtained from the local air quality control regions. Two CO levels were used from each area: an average for sites considered representative of the airshed and the highest average CO level. An overall total particulate emission rate of 0.48 g/mi was adopted for light diesel vehicles.

For each city, the average CO emission rate and the fraction of CO
attributable to light vehicles were calculated from the fleet
composition data of Abrott et al. (1978). In the four cities,
automobiles accounted for 70 to 80 percent of the ambient CO, with
emission rates of 59 to 65 g/mi in 1975.

Table 3.1 shows the calculated increases in ambient aerosol mass
loadings resulting from three levels of dieselization. On a mass
basis, the increase is less than 10 percent, except for certain sites
in Manhattan. The contribution to the submicron component of the
aerosol will be greater in general--perhaps by a factor of two--than to
the total.

Average increases in atmospheric benzo[a]pyrene concentrations
resulting from 25 percent dieselization of the 1975 light-duty vehicle
fleets in Manhattan and Los Angeles are 0.073 and 0.076 $\mu g/m^3$,
respectively. For Los Angeles this is a 17 percent increase over the
ambient benzo[a]pyrene concentrations of 0.46 $\mu g/m^3$ measured in
1975 (Abrott et al., 1978). In Manhattan, higher ambient
concentrations (1.4 $\mu g/m^3$) are reported (Kniep et al., 1979), and a
25 percent light diesel contribution represents a 5 percent increase.
In these calculations the spark-ignition engine's contribution to
benzo[a]pyrene levels was neglected because its measured emission rates
are 10 to 60 times smaller than those of diesels (Forrest et al., 1979;
Springer, 1978). An average benzo[a]pyrene emission rate of about 3.5
$\mu g/mi$ was taken from Oldsmobile and Volkswagen diesel emission rate
data (Springer, 1978).

VISIBILITY

Optical effects of aerosol particles, such as reduced visual range
or atmospheric heating, are proportional to the optical scattering and
absorption coefficients of the particles. Particle scattering, b_{sp},
has been shown to be correlated with particle mass in particle sizes
below 2.5 μm diameter, with a correlation coefficient typically of
0.95 and a ratio of scattering to mass of 1.3 to 4 m^2/g (Waggoner and
Weiss, 1980). Absorption, b_{ap}, is 5 to 7 m^2/g (Heisler et al.,
1980). The total light extinction caused by diesel-emitted material is
approximately 8 m^2/g.

The soot particles emitted by diesel engines absorb visible light.
Decreases in visibility (and possibly microclimate alterations) will
result from particle-induced changes in the atmospheric scattering and
absorption coefficients. Two calculations of diesel impacts on
visibility are presented below.

Visibility Reduction from Projected Primary Exhaust

The sum of absorption and scattering coefficients for both gaseous
and particulate emissions can be used to calculate visual range
(Middleton, 1952). Pierson (1978) reports that the extinction
coefficient, b (m^{-1}), per unit concentration ($\mu g/m^3$) of diesel

TABLE 3.1 Increase in Particulate Loading ($\mu g/m^3$) as a Function of Dieselization for the 1975 Fleet.

	Phoenix		Los Angeles		St. Louis		Manhattan		
	Typical	High	Typical	High	Typical	High	Typical	High	Hig[h]
Ambient (1975)	121		108		75		67		
Diesel- ization									
10%	1.0	1.7	2.5	2.9	1.4	3.1	2.4	6.1	13.
15%	1.5	2.6	3.8	4.3	2.1	4.6	3.6	9.1	20.
20%	2.5	4.3	6.3	7.2	3.5	7.7	6.0	15.2	33.

Particle loadings are calculated from equation (1) using ambient CO concentrations for these cases:

Typical - represents CO levels obtained from monitoring sites representative of the urban area.

High   - represents highest annual average CO concentration reported from a monitoring site in each urban area.

Highest - due to extreme variability in CO levels in Manhattan, the two highest annual averages for 1975 were used.

(Source: Local air quality offices; Phoenix, St. Louis, and Manhattan --Forrest et al., 1979; Los Angeles--Davidson et al., 1979)

aerosol emitted by trucks is $(b/\rho)_d = 8 \ m^2/g$. The effect of dieselization on visibility in Los Angeles can be estimated using this value for diesel exhaust under ambient conditions. Loadings caused by diesel exhaust are taken from the CO tracer calculations shown in Table 3.1. The visual range was calculated from the following expressions:

$$b_{ext} = \frac{3.9}{(\text{visual range})_{base\ case}} + \left(\frac{b_{ext}}{\rho}\right)_d \ (\text{diesel contribution to ambient aerosol g/m}^3) \quad (2)$$

and

$$\left(\frac{b_{ext}}{\rho}\right)_d = 8 \ m^2/g,$$

where b_{ext} is the sum of the absorption and scattering coefficients for both gaseous and particulate emissions. The results, shown in Table 3.2, are for the hypothetical cases in which 10 to 25 percent of the 1975 light-duty gasoline-powered fleet is replaced by diesels. Calculations indicate that significant reductions in visual range of up to 20 percent would accompany a 25 percent dieselization in Los Angeles.

The major uncertainties in these estimates are (1) the atmospheric concentrations of diesel particulate matter and (2) the extinction coefficient for diesel emissions in the atmosphere. The atmospheric concentrations are based on the use of the CO model (Eq. 1) and a particulate emission rate of 0.48 g/mi. The extinction coefficient assumed for diesel aerosol is about 8 m^2/g--a conservative figure.

Particle Extinction Measurements and Projections

The potential effects on visibility caused by changes in the fleet of light vehicles can also be made using measurements or estimates of present diesel and gasoline vehicle effects, projecting emissions, and calculating the change in ambient scattering and absorption. Measurements have been made of scattering (b_{sp}), using integrating nephelometer, and absorption (b_{ap}), using the integrating plate method on nuclepore filter particle samples, in many urban, rural, and remote locations. Values for b_{sp} range from $10^{-7}m^{-1}$ to 3 x $10^{-3}m^{-1}$; b_{ap} ranges from 5 x $10^{-8}m^{-1}$ to 2 x $10^{-4}m^{-1}$ (Weiss, 1980). Based on measurements made in Denver, Phoenix, Seattle, and Portland, the ratio of absorption to scattering is estimated at 0.3 to 1.0. Interpreting the same data in terms of visual range, a 25 to 50 percent reduction of urban visibility is caused by graphitic carbon (Weiss, 1980). The rural aerosol has a lower ratio of absorption to scattering, and both scattering and absorption are lower for rural aerosols.

Few definitive assignments of sources have been made for graphitic carbon in urban areas. Hansen et al. (1978) for San Francisco and Heisler (1980) for Denver estimate that all diesel vehicles are the source of about 35 percent of ambient graphitic carbon. In the most sophisticated source assignment, Wolff et al. (1980), using local source measurements and the C^{14}/C^{12} tracer technique, determined the following graphitic carbon source percentages for Denver in November and December of 1978: 39 percent wood burning, 20 percent diesel trucks, 5.3 percent gasoline cars and trucks (noncatalyst), 1.5 percent gasoline cars and trucks (catalyst), and 34 percent other sources, predominantly fuels burned for heating (primarily natural gas and fuel oil).

During the same period in Denver, Weiss (1980) found scattering and absorption approximately equal in magnitude. Projections can be made for two possible future scenarios starting with the source data of Wolff et al. (1980). The first would have no changes except that spark-ignition vehicles not equipped with catalytic converters (assumed to be half of the light vehicle fleet) are replaced by vehicles with catalytic converters, and the existing diesel emissions are decreased

TABLE 3.2 Visibility Estimates for Los Angeles.

Dieselization	Diesel Contribution to Ambient Aerosol $(\mu g/m^3)$ *	Overall Visual Range (Miles)	% Decrease in Visibility
1978 Base Case**	Negligible	10.5	--
10%	2.9	9.5	9.1
15%	4.3	9.1	13.0
25%	7.2	8.4	20.0

*   Represents "high values" taken from Table 3.1.

** Median value according to South Coast Air Quality Management District, Los Angeles.

to one-half of current levels. This results in a 14 percent decrease in graphitic carbon emissions and a 7 percent improvement in visibility over current levels (since scattering is unchanged). In a second projection, the number of diesel vehicles increase by a factor of 20, replacing 25 percent of the fleet; all other assumptions are unchanged. New diesel vehicles are assumed to emit only half the exhaust per vehicle mile of the current fleet. For this scenario, absorption in Denver would triple, and visual range would decrease to about one-half its current value.

Non-Urban Impact

Diesel particulate emissions remain in the atmosphere for as long as one week and can be transported distances on the order of 1,000 km. Resulting changes in the concentrations of graphitic carbon aerosols and soluble organic aerosols can be predicted using a simple flow-through or steady-state model.

Graphitic carbon is emitted from sources throughout the continental United States and is assumed to remain in the atmosphere long enough to be mixed. Neglecting to account for any removal mechanisms, the emitted carbon is all transported across the East Coast, and concentrations are calculated from source strengths and the flow volume of air. The flow is calculated assuming the length of the eastern seaboard to be 3,000 km, the vertical mixing height to be 3,000 m, and the average wind speed to be 5 meters per second; this yields 4.5 x 10^{10} m^3/s. The resulting carbon loading, based on EPA's "Best Estimate Particulate Control" case, is 0.3 $\mu g/m^3$ (U.S. EPA, 1980).

The national inventory of fine graphitic carbon particles was estimated from the National Emissions Data Summary for 1975 by halving the nonvehicular portion to account for fine particles only and taking 13 percent of the fine particle loading to obtain the graphitic carbon portion (Heisler et al., 1978). The vehicular portion was added from values provided by the EPA. Holding the nonvehicular portion fixed from 1975 to 2000 is equivalent to assuming that the use of particulate controls on stationary sources would be counterbalanced by the growth in emissions that are likely to occur because of conversions from oil to coal. This method gives 0.86 $\mu g/m^3$ for 1975 conditions and 1.1 $\mu g/m^3$ for the year 2000. Current rural graphitic carbon loadings, by comparison, are 1 $\mu g/m^3$ for rural Arkansas, 0.2 $\mu g/^3$ for Mesa Verde, Colorado, and 1 $\mu g/m^3$ for rural western Virginia (Weiss, 1980).

For all cases the reduction in rural visual range would be small, usually less than 3 percent, considering that only 10 percent of the optical extinction is caused by graphitic carbon and that one-third of this comes from diesels. Note that this Box model calculation is for uniform area emissions. Graphitic carbon from diesels would contribute much more than 0.3 $\mu g/m^3$ to the air in urban downwind plumes. A similar calculation for extractable organic material shows an impact of only 1 to 3 percent from diesels.

ATMOSPHERIC HEATING EFFECTS

Airborne graphitic particles absorb solar radiation, resulting in an increase in atmospheric heating and a decrease of heating at the earth's surface. The net climatic effect, calculated through models of the radiation transfer processes, is small—on the order of 0.1°C per day (Mitchell, 1971). The instantaneous rate of heating of air aloft caused by existing urban aerosols in St. Louis and Denver is as much as 5°C per day (Ackerman et al., 1976; Roach, 1961; Venkatram and Viskanta, 1977; Weiss, 1980). Detailed models, including evaporation and convective transport, predict thermal heating rates from haze to be 0.5°C per day and 1.5°C for 5 days. The heating aloft and reduced surface heating will act together to reduce mixing heights and increase the concentrations of all pollutants emitted at the surface. The heating effects are directly proportional to the total aerosol absorption.

The relative changes in urban CO concentrations with dieselization have been estimated by Forrest et al. (1979) for the case corresponding to the spark-ignition fleet anticipated in the year 2000. For Manhattan, St. Louis, and Phoenix, the CO emission rates from light vehicles would decrease by 20 percent with 25 percent dieselization. However, the larger contributions from heavy vehicles and stationary sources would cause the overall reduction from present CO levels to be 0.5 percent for Manhattan, 0.4 percent for St. Louis, but 4 percent for Phoenix.

Light vehicles accounted for 0.7 percent of the nation's sulfur dioxide (SO_2) emissions in 1975. Fractional contributions in the urban regions vary from 0.1 percent in St. Louis to 4 percent in greater Los Angeles. If 25 percent of the passenger cars and lightweight trucks were diesels, assuming that diesel engines emit 10 times as much SO_2 as spark-ignition engines, SO_2 emissions would be larger by 0.3 percent in St. Louis and 13 percent in Los Angeles. However, the importance of these emissions for secondary sulfate formation near roadways could be greater than the numbers indicate, because SO_2 from automobiles is emitted at ground level rather than from tall smokestacks.

CHEMICAL AND PHYSICAL TRANSFORMATIONS IN THE ATMOSPHERE

A comprehensive assessment of the health and environmental impacts of diesel exhaust must take into account the chemical, physical, and biological modifications of the emissions during transport through the atmosphere. The two principal concerns for diesel exhaust emissions in the atmosphere are (1) the capacity of primary gaseous pollutants to form secondary gaseous and particulate pollutants as photochemical smog and (2) the potential for significant modifications of polycyclic organic matter emissions by interactions with reactive molecular and free radical species.

Nitrogen Oxides

Diesel engines produce somewhat higher emissions of NO and NO_2 than spark-ignition engines fitted with catalytic converters (Andon et al., 1979). This is of concern because of the role these emissions play as precursors for the formation not only of ozone (O_3) but also the secondary nitrogenous pollutants. Photolysis of NO_2 remains the only known chemical pathway by which O_3 is formed in the atmosphere. NO and especially NO_2 also serve as precursors for the formation of a host of toxic or potentially toxic gaseous compounds.

Additional burdens of NO_x from diesel emissions should also be considered in terms of their impact on a phenomenon known as "acid rain," which in some parts of the western United States may be influenced by nitric acid (HNO_3) formed in the atmospheric reaction of NO_2 with the hydroxyl radical and ozone. Increased NO_x emissions will make it more difficult to achieve the federal air quality standard for NO_2 in those airsheds presently in violation of the standard.

Photochemical Smog Reactivity of Diesel Exhaust

Diesel exhaust contains substantially more aldehydes than gasoline engine exhaust. Aldehydes are known to catalyze the formation of photochemical smog and to irritate human lungs. In calculating diesel exhaust reactivity indices based on four criteria, Spicer and Levy (1975) found that while aldehydes represented about 30 percent of diesel hydrocarbon exhaust components on a volume basis, they accounted for 49 to 55 percent of the exhaust reactivity in all but the aerosol index.

Only a few studies of the smog-forming potential of diesel exhaust have been carried out (Dimitriades and Carroll, 1971; Landen and Perez, 1974; Spicer and Levy, 1975; Anderson and Hanley, 1980). Unfortunately, both calculated and experimental determinations of diesel exhaust reactivity have suffered from serious limitations, and the data derived from these studies are conflicting.

Sulfur Dioxide and Sulfate Aerosol Formation

Increased particulate carbon coupled with higher SO_2 emissions from diesels may contribute to greater ambient concentrations of secondary sulfate by catalyzing the heterogeneous oxidation of SO_2. The contribution of the homogeneous oxidation of SO_2 via attack by the hydroxyl radical (OH) and possibly the Criegee intermediate ($RCHO_2$, where R is an alkyl group) is unlikely to be substantially changed with increased dieselization. Ambient sulfate concentrations will increase with increases in diesel-powered light-duty vehicles.

Secondary Organic Aerosol Formation

In the absence of data gathered specifically for diesel exhaust, speculative considerations based on current knowledge concerning organic aerosol formation processes leads to a number of tentative observations and conclusions.

The distribution of diesel exhaust hydrocarbons is similar in composition to that of gasoline engine exhaust hydrocarbons but is shifted toward higher carbon numbers. This difference and the presence of carbonyl compounds may result in an increase in secondary organic aerosol formation, caused by the formation of lower volatility products. The magnitude of this aerosol effect, in relation to primary particulate carbon levels, remains to be determined. NO_x emissions in diesel exhaust may increase ambient levels of secondary inorganic nitrate as well as those of secondary nitrogen-containing organic compounds (aliphatic nitrate esters and aromatic nitro derivatives).

Transformations of Polycyclic Hydrocarbons on Diesel Particles

A growing body of evidence suggests that chemical and photochemical reactions may be a significant degradation pathway for adsorbed polycyclic aromatic hydrocarbons and aza-arenes in the atmosphere. Because the experimental evidence is limited, however, data from relevant model systems must be extrapolated or used as analogies. In view of the disparate and sometimes conflicting nature of the results obtained so far, manipulations of such models need to be performed with great caution. Clearly, the reactivity of adsorbed polycyclic aromatic hydrocarbons is an important subject for further investigation. The gaseous pollutants or atmospheric species that seem most likely to react chemically with polycyclic aromatic hydrocarbons in the polycyclic organic matter from diesel emissions are nitrogen dioxide, ozone, singlet molecular oxygen, and hydroxyl radicals.

Only limited information is available about the reaction products of the degradation of polycyclic aromatic hydrocarbons and their biological activities. They may or may not be more hazardous to humans than some of the original polycyclic aromatic hydrocarbons. Certainly, their polarity is commonly increased by chemical transformations in the atmosphere. This may lead in turn to greater biological activity.

A number of recent studies have treated the problem of polycyclic aromatic hydrocarbon degradation and transformation during filter collection (Lee et al., 1980; Pitts et al., 1978, 1980a, 1980b, 1980c; Schuetzle et al., 1980). However, distortions of the filtrate compositions may occur as a result of interactions with atmospheric pollutants and free radicals. Obtaining meaningful data on the atmospheric levels of diesel-derived polycyclic organic matter will depend on successful suppression or elimination of these artifacts.

ATMOSPHERIC TRANSPORT AND DEPOSITION

Various useful models for simulating the effects of increased diesel use on environmental quality have been developed. Methods are available for estimating the behavior of diesel soot as it is transformed and dispersed in the atmosphere. Such models could be extended to describe the ultimate fate of particulate emissions as they affect ecology and human exposure. The following section addresses some of the issues surrounding available methods for estimating pollutant concentrations.

If diesel particles were a pure substance whose vapor pressure were 10^{-9} to 10^{-8} torr, it would take days or weeks for them to vaporize, according to estimates used in standard formulas (Hidy and Brock, 1970). The rate of attack of polycyclic aromatic hydrocarbons by hydroxyl radicals in the air gives half-lives in the gas phase that are short compared with evaporation times. As the material volatilizes it rapidly degrades. If it does not volatilize, its lifetime in the atmosphere is set by fine particle removal mechanisms such as washout and dry deposition. Studies by Moore et al. (1973) suggest that diesel particles are airborne at lower altitudes about 5 days and in the upper troposphere 10 to 15 days. A conservative estimate of particle lifetime is a week to a month; if evaporation and chemical degradation intervene, ambient concentrations will be decreased below these levels. This estimate permits the use of nonreactive modeling to determine expected particle concentrations.

A wide variety of mathematical models are available for a systems analysis of the behavior of diesel soot in the atmosphere. Simulation models can be used to link air quality and emissions patterns in a specific meteorological setting. The models consider a wide range of geographical and temporal factors. Roadway corridor models are available for estimating concentration distributions for mobile source pollutants in the presence of steady wind and meteorologically stable conditions. Urban population exposures can be assessed using numerical grid or multisource Gaussian formulations. The long residence times of diesel aerosols suggest the use of one of the numerous long-range transport models. Special purpose packages are available for computing visibility, chemical transformations, and emissions dispersion-transport processes.

The Denver Winter Haze Study, for example, found close agreement between measured atmospheric concentrations of various pollutants, including organic and inorganic carbon, and pollutant concentrations calculated from measured emission factors, fuel usage in mobile and stationary sources, and dispersion (Heisler et al., 1980b; Wolff et al., 1980). In Denver, primary organic and inorganic aerosols were the dominant sources of degraded visibility; secondary organics, sulfate, and nitrate played smaller roles. As discussed in the section on chemical transformations, both gas phase and surface processes require additional research before they can be described in transport and deposition models.

In addition to atmospheric considerations, the question of the ultimate fate of the material needs to be addressed after it is deposited on the surface of soil or on a body of water. A natural extension of atmospheric modeling and deposition research involves tracing pathways of the materials through the soil, groundwater, and surface water. Additional processes (such as hydrolysis, photolysis, and biodegradation) that influence the fates of these pollutants must be considered in this assessment. Such pathways may be important for determining both ecological impacts and human exposures by way of drinking water, food, or skin contact.

CONTRIBUTION OF DIESEL EMISSIONS TO ATMOSPHERIC AEROSOLS

Although it is possible to estimate the effects of light-duty diesel use on air quality by using CO emission rates as indicators of particulate dilution rates, identifying a signature or tracer characteristic of diesel particles would be useful for determining the contribution of diesels to atmospheric aerosols. Examination of roadway data (Pierson and Bracheczek, 1976; Dzuboy et al., 1979) and laboratory data (Hare and Baines, 1979; Hare and Bradow, 1979) show that, aside from the major constituents of the particles (carbon, hydrogen, oxygen, nitrogen, and sulfur), the only elements that appear consistently are phosphorus, zinc, calcium, and barium. Preliminary examination of the data shows no clear correlation among these elements, but further examination of their interrelationships is probably justified. A detailed size-classified elemental analysis of diesel particles might yield a clearer pattern. There has been no significant research specifically aimed at identifying a telltale signature for diesel particulates.

One possible method of identifying the contribution of diesel exhaust is through the use of an organic tracer. None has yet been demonstrated, though some preliminary data on potential organic tracers are available. The high molecular-weight solvent-extractable fraction, which is adsorbed on or entrapped in the black diesel soot, has been discussed as the most promising fraction for investigation (e.g., Simoneit et al., 1980). Diesel exhaust has a strong and characteristic odor, so the human sense of smell can be a good detector of gross contamination. The smoky odor characteristic of diesel exhaust has been ascribed to hydroxy- and methoxyindanones, with some contribution from methyl- and methoxyphenols; burnt odors have been associated with furans and alkylbenzaldehydes (Levins et al., 1974).

Data on the organic components of diesel exhaust in ambient aerosols are sparse. A study of nonregulated emissions from vehicles, being carried out by Ford Motor Company, has provided some preliminary details (Gorse, 1980). The carbon number range of the gaseous compounds is from about C_5 to C_{24}, and the classes consist of various aliphatic and aromatic hydrocarbons. The hydrocarbons in the particulate emissions (CH_2Cl_2 extractable) ranged from C_{14} to about C_{40}, with no carbon number predominant and a concentration maximum at about C_{26}. The extracted polar material had a polycyclic aromatic hydrocarbon distribution that was apparently different for gasoline and diesel engine emissions, with the latter having a dominance of lower molecular weight material.

Residual inorganic carbon, which is more commonly called soot, makes up about 72 percent of the carbon in diesel emissions (Gorse, 1980). It is the predominant form of carbonaceous material in urban aerosols (Grosjean et al., 1980; Rosen et al., 1980). Carbonaceous soot is emitted from all combustion processes because a limited availability of oxygen prevents the complete conversion of carbon to CO_2. Accordingly, it is difficult to specifically identify the source of soot in the ambient environment.

COMPARISON OF DIESEL EMISSIONS WITH AMBIENT AEROSOLS

Most chemical analyses of the organic constituents of diesel exhaust and their concentrations have been done on the volatile fraction. Only preliminary examinations have been undertaken of the solvent extractable material and the residual organic carbon. The volatile compounds in ambient urban aerosols come from various sources. The individual contributors are difficult to identify because all sources, including diesel, emit very similar materials. The only groups of compounds not found in vehicular exhaust are the plant terpenes, but their periods in the atmosphere are short because of their high reactivity. The solvent-extractable material has the greatest potential for being characteristic of vehicular exhausts.

ECOLOGICAL EFFECTS

Besides graphitic carbon particles, diesel exhaust contains many organic compounds that may be adsorbed on single and agglomerated particles. Depending on concentration, some of these compounds have been found to be toxic to microbes and to induce biochemical and physiological changes in the lung tissues of rodents (see Chapter 4). Of particular concern are some polycyclic aromatic hydrocarbons, such as benzo[a]pyrene, which are sufficiently long-lived in oil spills and sediments to accumulate over time.

The polycyclic aromatic hydrocarbons observed in ecosystems come from both natural and anthropogenic sources. The latter dominate, especially in and near densely populated areas. Surface capture may be an important mechanism for exposure of terrestrial organisms to particulates. However, absorption and assimilation appear to vary a great deal. The potential exists for at least limited biomagnification for some substances along food chains.

Large amounts of polycyclic aromatic hydrocarbons have been released in aquatic systems by oil spills, so that some information about their effects on water-living organisms is available. It is known that in high concentrations many polycyclic aromatic hydrocarbons are toxic to aquatic animals, inhibiting growth, interfering with embryonic development, altering physiology and behavior, and producing tumors. However, at present we can do little more than speculate about the sensitivities of organisms and communities to diesel particulates.

It appears unlikely that a substantial increase in diesel emissions will perceptibly affect ecosystems distant from the sources of emissions. Research on potentially sensitive organisms and communities within these ecosystems is lacking, particularly on the deposition and accumulation of diesel particles over long periods of time.

HEALTH EFFECTS

Previous studies of human health effects associated with diesels have not usually addressed the question of the effects from secondary substances formed when sunlight irradiates diesel exhaust in the air. There are many difficulties in performing such studies. Epidemiologic studies have emphasized occupational groups, such as coal miners and bus mechanics. Deliberate exposure of human subjects to irradiated diesel exhaust is not possible. Moreover, diesel inhalation studies of animals have seldom included irradiation. Nevertheless, some potentially important effects are suspected. A detailed discussion of research efforts on the mutagenic, carcinogenic, and pulmonary and systemic effects of diesel exhaust exposure is presented in Chapter 4.

In vitro assays have indicated that directly active mutagens are formed when benzo[a]pyrene is exposed to gaseous pollutants in photochemical smog (Pitts et al., 1978). Diesel exhaust reaction products may irritate the respiratory tract or compromise the human defenses against respiratory infection (Campbell et al., 1980) or both. Greater numbers of diesels in localities already experiencing photochemical smog may increase the conversion of gaseous sulfur oxides to sulfur-containing aerosols. This possibility presents a health concern because high levels of SO_2 and coexisting particulate pollutants have been associated repeatedly with increases in respiratory morbidity and mortality rates (NAS, 1977; NAS, 1978).

HEALTH RESEARCH AGENDA

Providing the needed information on human health effects of irradiated diesel exhaust will require a wide range of experimental subjects and investigative tools. Animal and _in vitro_ studies generally employ large pollutant doses, but investigations at lower, more "realistic" doses should be encouraged where practical to minimize the need for speculative extrapolations from high doses for predicting ambient exposure effects. Laboratory studies often require extensive resources such as large smog chambers, animal exposure facilities, and animal testing equipment, while epidemiologic studies generally require large populations. To do such studies, substantial funding commitments will be needed in all areas of investigation.

Many epidemiologic and toxicologic studies of photochemical oxidants, sulfur oxides, and nitrogen oxides are in progress or being planned. Some may be directly relevant to diesel risk assessment. Others may be made useful by appropriate modification or addition to a study. Epidemiologic studies of ambient exposure generally concentrate on the areas where the problem under study is most severe, such as the examinations of photochemical oxidants in Los Angeles. It may be informative to perform occupational studies of outdoor workers exposed to diesel exhaust in the same harshly polluted areas.

Most studies of organic particulate pollutants, including reaction products of polycyclic aromatic compounds, have been primarily concerned with documenting their behavior in the atmosphere, either in ambient air or in smog chambers. More attention should be given to possibilities for combining health studies with atmospheric studies. The relevance of research on organics and other poorly investigated categories of pollutants, such as SO_2 oxygenation in the presence of diesel soot, could be increased by placing more emphasis on animal inhalation toxicology.

Among the atmospheric contaminants expected to increase in concentration with increased dieselization are reaction products of polycyclic aromatic hydrocarbons, secondary organic aerosols, and sulfur-containing aerosols. Of these, only sulfur-containing aerosols have been studied extensively. Epidemiologic studies have suggested that some human respiratory problems are associated with sulfate pollution. This has not been confirmed in laboratory studies of human volunteers. However, complex systems with SO_2 oxidation in the presence of soot particles have seldom been included in such studies. Only limited health effects information is available, mostly from _in vitro_ studies, on other diesel-generated pollutants. Carcinogenic or mutagenic potential has been demonstrated for some substances, but the public health implications of these findings are unclear at present. Extensive studies of carcinogenesis and non-neoplastic respiratory and systemic effects are needed to evaluate the health effects of diesel-contaminated air.

ENVIRONMENTAL RESEARCH AGENDA

Emissions Characterization

* Adsorption kinetics influence gas-to-particle transfer
  processes under roadway conditions where the general public is
  most likely to be exposed. The partitioning of adsorbable
  materials between the gas and particle phases may have an
  important influence on their environmental effects.
  Therefore, research is needed on the factors governing
  partitioning, especially adsorption.

Air Quality Effects of Primary Emissions

* Preliminary calculations show that increased use of diesel
  engines will cause significant reductions in visibility in
  certain urban air sheds. Such calculations have been based on
  the optical properties of diesel aerosols and the use of CO as
  a tracer for automobile emissions. Additional calculations
  should be made using more recent data and alternative models.

* Airborne graphitic particles absorb solar radiation, resulting
  in an increase in atmospheric heating and reduced heating at
  the earth's surface. Research is needed on the potential net
  climatic effect and increased surface pollutant concentrations
  caused by increases in diesel-generated particulates.

Chemical and Physical Transformations in the Atmosphere

* The distribution of diesel hydrocarbons is similar in
  composition to that of gasoline engine exhaust hydrocarbons
  but is shifted toward higher carbon numbers. Experimental
  research should be initiated, after careful evaluation, to
  determine if this difference and the presence of carbonyl
  compounds results in an increase in secondary organic aerosol
  formation and what would be the magnitude of such an effect.

* Direct experimental investigation of the reactivity of diesel
  exhaust is urgently needed, using the most current analytical
  and smog chamber methods. To overcome any limitations of
  previous studies, it may be necessary to use a synthetic
  diesel exhaust to simulate the release of emissions into
  atmospheres with appropriate HC/NO_x ratios. With sufficient
  periods of light irradiation exposures, the full chemical
  reactivity of the individual organic constituents in diesel
  exhaust will be realized.

* Observations of filter artifacts and reactions of polycyclic
  aromatic hydrocarbons on other substrates raise the question

of whether polycyclic aromatic hydrocarbons adsorbed on the surface of airborne diesel particles react similarly in the atmosphere. Research is needed to determine the influence of surface chemistry, as well as the effects of pollutant levels, particle sizes, sunlight intensity, atmospheric mixing, and transport time on the atmospheric reactions of polycyclic aromatic hydrocarbons.

- The kinetics and mechanisms of the thermal and photochemical transformations of selected polycyclic aromatic hydrocarbons in diesel polycyclic organic matter should be investigated with single gaseous pollutants as well as with ambient photochemical smog. Exposure studies of model heterogeneous systems, as well as real particulate matter for diesel exhaust, should be performed in environmental chambers under controlled conditions.

Atmospheric Transport and Deposition

- Although some estimates of particulate levels in the continental plume are given by the committee, research is needed to provide a model of atmospheric transport and deposition of diesel-emitted particulates.

- In addition, urban grid model simulations and plume calculations should be employed to determine the urban time-space distributions of various pollutants. Existing data banks for baseline cases can be used, along with emissions inventories modified for various potential diesel penetrations of the light vehicle fleet. Exposures could then be estimated for fine particles, ozone, and nitrogen dioxide at various levels of dieselization.

Diesel Contribution to Atmospheric Aerosols

- Research is needed to establish thoroughly the molecular composition of diesel and gasoline exhaust. Organic composition data for other sources, especially for the natural background, should be gathered to aid in defining more precisely the contribution of diesel emissions to ambient urban and rural atmospheres. Secondary aerosol reactions, as well as the health hazards, can then be evaluated. Research should be aimed at identifying a diesel particulate signature or tracer in the air.

- Source resolution methods have been developed for relating air quality to particulate emissions. Such methods have relied on subtracting the chemical components of known sources from the aerosol. The greatest uncertainty remains in establishing the sources of the carbonaceous component of the aerosol, including the diesel contribution. While a few studies in this important field have been carried out, much more work is needed.

• A continuing air monitoring program should be initiated in selected urban areas to determine the levels of diesel components of the aerosol. Such components should include soot, certain polycyclic aromatic hydrocarbons, and the optical absorption coefficient of the aerosol.

Ecological Impacts of Diesel Emissions

• Research is needed on potentially sensitive organisms and communities within ecosystems distant from the source of diesel engine emissions, particularly on the deposition and accumulation of diesel particles over long periods of time.

Health Effects of Diesel Exhaust Reaction Products

• A wide range of experimental subjects and investigative tools are needed to provide information on the human health effects of irradiated diesel exhaust. To facilitate this, substantial commitments are needed in many approaches of health effects research, including smog chambers, animal exposure facilities, animal testing equipment, and large-scale epidemiological investigation of diesel exhaust emissions.

REFERENCES

Abrott, T. J., M. J. Barlelona, W. H. White, S. K. Friedlander, and J. J. Jorgan (1978). Human Dosage/Emission Source Relationships for Benzo[a]Pyrene and Chloroform in the Los Angeles Basin. EPA Contract 68-03-0434.

Ackerman, T. P., K. N. Lion, and C. B. Leavy (1975). Infrared Radiative Transfer in Polluted Atmospheres. Applied Meteorology 15 28-35.

Anderson, R. J., and J. T. Hanley (1980). A Smog Chamber Study of Aerosol Formation and Growth Involving SO_2 and Diesel Exhaust. Final Report, EPA Contract 68-02-2987. Buffalo, NY: Calspan Corporation.

Andon, J., et al. (1979). An Initial Assessment of the Literature on the Measurement, Control, Transport, Transformation and Health Effects of Unregulated Diesel Engine Emissions. South Coast Technology, Inc. Santa Barbara, CA. Prepared for US DOT-NHTSA. Publication DOT HS-804 010.

Black, F., and L. High (1979). Methodology for Determining Particulate and Gaseous Diesel Hydrocarbon Emissions of Particulate and Gas Phase. SAE Technical Paper 79-0422.

Bradow, R., F. Black, P. Gabele, and J. Ma (1979). Sampling Diesel Particles. SAE Technical Paper 790423.

Campbell, K. I., E. L. George, and I. S. Washington, Jr. (1980). Enhanced Susceptibility to Infection in Mice after Exposure to Dilute Exhaust from Light Duty Diesel Engines. Health Effects of Diesel Engine Emissions: Proceedings of an International Symposium. Volume 2. EPA-600/9-80-057b.

Davidson, A., M. C. Hoggan, and M. A. Nazemi (1979). Air Quality Trends in South Coast Air Basin. South Coast Air Quality Mangement District, El Monte, CA.

Dimitriades, B., and H. B. Carroll, Jr. (1971). Photochemical Reactivity of Diesel Exhaust--Some Preliminary Tests. Bureau of Mines Report of Investigation 7514.

Dzuboy, T.G., R.K. Stevens, W. J. Courtney, and E.A. Drone (1979). Chemical Element Balance Analysis of Denvery Aerosol. Paper presented at Symposium on Electron Microscopy and X-Ray Applications to Environmental and Occupational Health Analysis. September 1979. Colorado Springs, CO.

Forrest, L., W. B. Lee, and W. M. Smalley (1979). Impacts of Light Duty Vehicle Dieselizaton on Urban Air Quality. Paper presented at 72nd Annual Meeting of Air Pollution Control Association, June 1979. Cincinnati, OH.

Forrest, L., W. B. Lee, W. M. Smalley, and J.C. Sturm (1979). Impacts of Light-Duty Vehicle Dieselization on Urban Air Quality. Paper presented at the 72nd Annual Meeting of the Air Pollution Control Association, June 1979. Cincinnati, OH.

Frisch, L. E., J. H. Johnson, and D. G. Leddy (179). Effects of Fuels and Dilution Ratio on Diesel Particulate Emission. SAE Technical Paper 79-417.

Fuller, E. L., Jr., R. N. Caron, K. J. Fallon, J. F. Orrik, and J. L. Roux-Buisson (1978). Surface Properties of Wyodak Coal. Oak Ridge National Laboratory Publication ORNL/MIT-273.

Gorse, R. A., Jr. (1980). Ford Motor Company Nonregulated Emissions Characterization Program. Presentation to the Diesel Impacts Study Committee, Environmental Impacts Panel. May 1980. Los Angeles.

Gorse, Robert A., Jr. (1980). Ford Motor Company Nonregulated Emission Characterization Program. Detroit, MI: The Ford Motor Company.

Grosjean, D., and S. K. Friedlander (1975). Gas-Particle Distribution Factors for Organic and Other Pollutants in the Los Angeles Atmosphere. APCA Journal. 25:1038-1044.

Grosjean, D., K. Fung, P. Mueller, S. Heisler, and G. Hidy (1980). Particulate Organic Carbon in Urban Air: Concentrations, Size Distribution and Temporal Variations. AIChE Symposium Series AIR-1979 (in press).

Hansen, A. D. A., W. H. Brenner, and T. Novakov (1978). A Carbon and Lead Emission Inventory for the Greater San Francisco Bay Area. Atmospheric Aerols Research Annual Report, 1977-1978. Lawrence Berkeley Laboratory. Document LBL-8686.

Hare, C. T., and R. B. Bradow (1979). Characterization of Heavy-Duty Diesel Gaseous and Particulate Emissions and Effects of Fuel Composition. The Measure and Control of Diesel Particulate Emissions. SAE Publication PT-79/17.

Hare, C. T., and T. M. Baines (1979). Characterization of Particulate and Gaseous Emissions from Two Diesel Automobiles as Functions of Fuel and Driving Cycle. The Measurement and Control of Diesel Particulate Emissions. SAE Publication PT-79/17.

Heisler, S. L., R. C. Henry, J. G. Watson, and G. M. Hidy (1980a). The 1978 Denver Haze Study. Environmental Research and Technology, Inc., Document P-5417-1.

Heisler, S. L., R. C. Henry, J. G. Watson, and G. M. Hidy (1980b). The 1978 Denver Winter Haze Study. Environmental Research and Technology, Inc., Document P-5417-2.

Hidy, G. M., and J. R. Brock (1970). The Dynamics of Aerocolloidal Systems. London: Pergamon Press.

Japar, S. M., and Szkarlat, A. (1980). Mass Monitoring of Carbonaceous Aerosols by Spectrophone. Combustion Science and Technology. 234:143.

Johnson, W. B., D. E. Wolf, and R. L. Mancuso (1978). Long-Term Regional Patterns and Transfrontier Exchanges of Airborne Sulfur Pollution in Europe. Atmospheric Environment. 12:511-527.

Kittleson, D. B., and D. F. Dolan (1980). Diesel Exhaust Aerosols. In Proceedings of Symposium on Aerols Generation and Exposure Facilities. Klaus Wikkele (ed.). Ann Arbor, MI: Ann Arbor Scientific Press.

Kneip, T. J., M. Lippman, F. Mukai, and J. M. Daisey (1979). Trace Organic Compounds in the New York City Atmosphere Part I--Preliminary Studies. EPRI EA-1121, Project 1058-1.

Landen, F. W., and J. M. Perez (1974). Some Diesel Exhaust Reactivity Information Derived by Gas Chromatography. SAE Technical Paper 740530.

Levins, P. L., D. A. Kendall, A. B. Caragay, G. Leonardos, and J. E. Oberholtzer (1974). Chemical Analysis of Diesel Exhaust Odor Species. SAE Automotive Engineering Congress. Detroit, 985-995.

Lee, F. S. C., W. R. Pierson, and J. Ezike (1980). The Problem of PAH Degradation during Filter Collection of Airborne Particulates—An Evaluation of Several Commonly Used Filter Media. Proceedings of the Fourth International Symposium on Polynuclear Aromatic Hydrocarbons. October 1979. Columbus, OH: Battelle Publishing, Inc.

Lipkea, W. H., J. H. Johnson, and C. T. Vuk (1978). The Physical and Chemical Character of Diesel Particulate Emissions, Measurement Techniques and Fundamental Considerations. SAE Special Publication 430.

Lipkea, W. H., J. H. Johnson, and C. T. Vuk (1979). The Physical and Chemical Character of Diesel Particulate Emissions—Measurement Techniques and Fundamental Consideration. The Measurement and Control of Diesel Particulate Emissions. SAE Publication PT-79/17.

Lippmann, M. (1976). Size-Selective Sampling of Inhalation Hazard Evaluations. In Fine Particles. B. Y. H. Liu (ed.). New York: Academic Press.

MacDonald, J. S., S. L. Plee, J. B. C'Arcy, and R. M. Schreck (1980). Experimental Measurements of the Independent Effects of Dilution Ratio and Filter Temperature on Diesel Exhaust Particulate Samples.

Middleton, W. E. K. (1952). Vision Through the Atmosphere. Toronto: University of Toronto Press.

Mitchell, J. M., Jr. (1971). The Effect of Atmospheric Particles on Radiation and Temperature. In Man's Impact on Climate. W. H. Matthews, W. W. Kellogg, and G. O. Robinson (eds.). Cambridge, MA: The MIT Press.

Moore, H. E., S. E. Poet, and E. A. Martell (1973). ^{222}Rn, ^{210}Pb, ^{210}Bi, and ^{210}Po Profiles and Aerosol Residence Times versus Altitude. Journal of Geophysical Research, 78:7065-7075.

National Research Council (1977). Airborne Particles. Report by the Committee on Medical and Biological Effects on Environmental Pollutants. Washington, D.C.: National Academy of Sciences.

National Research Council (1978). Sulfur Oxides. Report by the Committee on Sulfur Oxides. Washington, D.C.: National Academy of Sciences.

Natusch, D. F. S., and B. A. Tomkins (1978). Theoretical Consideration of the Adsorption of Polynuclear Aromatic Hydrocarbon Vapor onto Fly Ash in a Coal-fired Power Plant. Carcinogenesis, Vol 3: Polyncuearl Aromatic Hydrocarbons, P. W. Jones and R. I. Freudenthal (eds.) New York: Raven Press.

Pierson, W. R. (1978). Particulate Organic Matter and Total Carbon from Vehicles n the Road. National Science Foundation-Lawrence Berkeley Laboratory Conference on Carbonaceous Particles in the Atmosphere, Berkeley, CA, March 20-22.

Pierson, W. R., and W. W. Brachaczek (1976). Particulate Matter Associated with Vehicles on the Road. SAE Technical Paper 760039.

Pitts, J. N., Jr., K. A. van Cauwenberghe, D. Grosjean, J. P. Schmid, D. Fitz, W. L. Belser, Jr., G. B. Knudson, and P. W. Hyunds (1978). Atmospheric Reactions of Polycyclic Aromatic Hydrocarbons: Facile Formation of Mutagenic Nitro Derivatives. Science 202(3):515-519.

Pitts, J. N., Jr., D. M. Lokensgard, A. M. Winer, G. W. Harris, and S.D. Shaffer (1980a). Photochemical and Thermal Reactions of Combustion-Related Particulate Organic Matter. Paper presented at meeting of the Physics and Chemistry of Energy-Related Atmospheric Pollution; A Program Review by the Office of Health and Environmental Research, U.S. Department of Energy, May 1980, Harpers Ferry, West Virginia.

Pitts, J. N., Jr., D. M. Lokensgard, P. S. Ripley, K. A. Van Cauwenberghe, L. Van Vaeck, S. D. Shaffer, A. J. Thill, and W. L. Belser, Jr. (1980b). "Atmospheric" Epoxidation of Benzo(a)pyrene by Ozone: Formation of the Metaboilte Benzo(a)pyrene-4,5-Oxide. Science (submitted for Publication).

Pitts, J. N., Jr., A. M. Winer, D. M. Lokensgard, S. D. Shaffer, E. C. Tuazon, and G. W. Harris (1980c). Interactions Between Diesel Emissions and Gaseous Co-Pollutants in Photochemical Air Pollution: Some Health Implications. Health Effects of Diesel Engine Emissions: Proceedings of an International Symposium. Volume 1. EPA-600/9-80-057a.

Roach, W. T. (1961). Some Aircraft Observations of Solar Radiation in the Atmosphere. Quarterly Journal of the Royal Meterological Society. 87:340-363.

Rosen, H., A. D. A. Hansen, R. L. Dod, and T. Novakov (1980). Soot in Urban Atmospheres: Determination by an Optical Absorption Technique. Science. 208:741-744.

Rothenberg, S. J., and Y. S. Cheng (1980). Coal Combustions Fly Ash Characterization: Rates of Adsorption and Desorption of Water. To be Published J. Phys. Chem.

Schuetzle, D., F. S. C. Lee, T. J. Prater, and S. B Tejada (1980). The Identification of Polynuclear Aromatic Hydrocarbon Derivatives in Mutagenic Fractions of Diesel Particulate Extracts. Paper presented at the 10th Annual Symposium on the Analytical Chemistry of Pollutants. May 1980. Dortmund, Germany.

Sherrer, H. C., D. B. Kittelson, and D. F. Dolan (1980). Light Scattering and Absorption by Diesel Aerosols, University of Minnesota Particle Laboratory Publication 418.

Simoneit, B. R. T., M. A. Mazurek, and T. A. Cahill (1980). Contamination of the Lake Tahoe Air Basin by High Molecular Weight Petroleum Residues. J. Air Poll. Control Assoc. 30:387-390.

Spicer, C. W., and A. Levy (1975). The Photochemical Smog Reactivities of Diesel Exhaust Organics. Report to the Coordinating Research Council. Columbus, OH: Battelle Columbus Laboratories.

Springer, K. J. (1978). Exhaust Particulate--The Diesel's Achilles Heel. Paper presented at 71st Annual Meeting of the Air Pollution Control Association, Jun 1978. Houston, TX.

U.S. Environmental Protection Agency (1980). Health Assessment Document for Particulate Organic Matter. External Review Draft No. 2, June 1980 (under revision). (We are indebted to Robert Bruce and Michael Berry for permission to quote from this draft document.)

Venkatram, A., and R. Viskanta (1977). Effects of Aerosol Induced Heating on Convective Boundary Layer. Journal of American Science. 34:1918-1933.

Waggoner, A. P., and R. E. Weiss (1980). Comparison of Fine Particle Mass Concentration and Light Scattering Extinction in Ambient Aerosol. Atmospheric Environment. 14:623-626.

Weiss, R. E. (1980). Optical Absorbtion Properties of Suspended Particles in the Lower Troposphere at Visible Wavelengths. Ph.D. Dissertation. Seattle, WA: University of Washington.

Whitby, K. T. (1978). The Physical Characteristic of Sulfur Aerosols. In Sulfur in the Atmopshere. New York: Pergamon Press, p. 135.

Williams. R. L., and C. R. Begeman (1979). Characterization of Exhaust Particulate Matter from Diesel Automobiles. Warren, MI: General Motors Research Laboratories. GMR-2970 and ENV-61.

Wolff, G. T., R. J. Couness, P. J. Groblicki, M. A. Ferman, S. H. Cadel, and J. L. Mulbaer (1980). Visibility Reducing Species in Denver Brown Cloud, Part II, Sources and Temporal Patterns. General Motors Research Laboratory Publication GMR-3391.

4 HEALTH EFFECTS

Trace amounts of carcinogenic and mutagenic substances are formed whenever organic matter is burned. Many of these substances belong to the class of compounds known as polycyclic aromatic hydrocarbons. The formation of these substances depends on the nature and composition of the original material and the conditions under which it is burned. For at least 25 years scientists have known that polycyclic aromatic hydrocarbons are present in the exhausts of both gasoline and diesel engines (Kotin et al., 1954a, 1954b; Hare and Baines, 1979). Diesel engines also produce 30 to 100 times the concentration of particulate produced by gasoline engines equipped with catalytic converters (Williams and Swarin, 1979). Consequently, the projected increase in diesel cars has raised concern about a significant rise in carcinogenic combustion products in the environment and their potential adverse effects on people.

The particles in diesel engine exhaust are miniscule, mainly in the submicron range. Because of their small size, they remain suspended in the air for a week or more and they can be inhaled and deposited in the narrowest passages of the lungs. Because of their small size, however, they offer a relatively large surface on which toxic, mutagenic, and carcinogenic compounds can be adsorbed.

Attention to carcinogenesis and mutagenesis centers primarily on particulate emissions, though equal consideration needs to be given to potential alterations to the ambient air caused by gases emitted in the diesel exhaust.

Such emissions can increase the ambient concentrations of certain pollutants, such as SO_x, NO_x, and volatile aldehydes (particularly formaldehyde and acrolein). Changes in the concentrations caused by the addition of diesel exhaust gases to the ambient air depend in part upon the nature of the fuel, the operating characteristics of the engine, and the available emission control system. The extent to which components of the gaseous portion would undergo atmospheric chemical reactions (as in photochemical smog) that create other toxic volatile substances (e.g., ozone and peroxacetyl nitrate) must also be considered as a potentially important contribution to altering the ambient air environment.

Information is needed about the levels of diesel exhaust constituents that humans are exposed to in work places and in such

65

ambient settings as city streets. Several environmental and
physiological factors may bear on the potential threat of diesel
exhaust to human health. Once emitted from the tailpipe, the exhaust
is subject to environmental dispersion, transport, and chemical
transformation--all capable of altering its components and
concentrations in the ambient air at the point of human contact.
Because diesel exhaust is one of many sources of air pollution, the
relevant issue is the risk to human health from the incremental
addition of diesel emission constituents to the existing state of the
ambient air.

As a consequence of this concern, national standards for some
gaseous pollutants, including oxidants such as CO, SO_2, and NO_2,
have been promulgated and others are under discussion. In view of the
questions being raised, the potential for elevated peak ambient levels
of NO_x (expressed as NO_2), for example, through increased diesel
emissions may be at least transiently sufficient to exceed proposed
standards, with possible harmful effects on susceptible populations.
Thus, it is reasonable to assume that increased levels of particulates
may aggravate pulmonary and systemic disfunction, because both the
small particulate size and secondary aerosol formation would enhance
the deposition of other toxic noncarcinogens deep in the lungs.

Although several federal agencies are now conducting research on
the potential health hazards of particulates and certain chemical
compounds in diesel exhaust, the existence of adverse health effects in
humans, including carcinogenesis, has not been demonstrated. Such
scientific uncertainty about health effects poses a dilemma for the
government policy-maker who must assess the health risks of exposure to
diesel engine exhaust.

The relationship between the concentrations at which effects have
been observed in laboratory animals or on bacterial cells and the
dosage to human lung tissue exposed to ambient concentrations is not
known. The rate of deposition of particulate matter and associated
chemical components in the human lung can be estimated from lung
deposition models and the chemical composition of the aerosol. It may
be difficult, however, to relate such lung deposition calculations to
the animal and cell exposure data. Analyses of this kind are needed
nevertheless to put such studies in perspective.

The health hazards of diesel exhaust have not been compared
comprehensively with emissions from gasoline-powered cars that are
likely to be replaced by diesels. Secondary exhaust products from
environmental transformations might be expected to differ
quantitatively because of their induction by higher levels of SO_2 and
NO_2 in diesel exhaust than in gasoline exhaust, as well as higher
levels of O_3 in the ambient air. An additional complication is the
diversity of diesel engines that already contribute to ambient
pollution levels--predominantly those in stationary sources, but also
in locomotives, trucks, tractors, buses, and, increasingly, passenger
automobiles.

MUTAGENESIS

A significant observation, made in a bacterial assay in 1978, showed that diesel exhaust particulates suspended in an organic solvent produced mutations. It demonstrated not only that the extractable materials could damage chromosomal deoxyribonucleic acid (DNA) but also that they might be carcinogenic (Huisingh et al., 1978). The latter conclusion is based on the observation that most chemical carcinogens have been shown to be mutagenic in a diverse group of submammalian and in vitro mammalian assays (Miller and Miller, 1971; Brusick, 1978; Magee, 1977; Ames et al., 1973; Bouck and di Mayorca, 1976; Ames, 1979; McCann et al., 1975).

Short-term submammalian and in vitro bioassays will be valuable tools to develop supporting data for in vivo carcinogenesis studies. Their predictive nature and operational flexibility make them ideally suited to answer questions not amenable to study by conventional in vivo bioassays. For example, studies comparing the biological activity of diesel exhaust particulate extracts prepared with a variety of organic solvents and biological fluids would be an impossible task using in vivo bioassays. They can be readily performed, however, with a wide range of short-term bioassays.

Different classes of chemicals produce different types of mutation and chromosomal damage in DNA. Therefore, several kinds of short-term tests need to be used in a comprehensive evaluation of diesel exhaust particulates and their extracts. These should include gene mutation, chromosomal aberhations, and DNA repair end-points.

In vivo studies address the issues of somatic cell and heritable genetic damage. Most important will be the analysis for induction of heritable changes in the gametes arising from inhalation exposure. The in vivo tests should also cover the major types of genetic end-points mentioned above.

The evidence for genotoxic activity of diesel exhaust and hence for carcinogenic potential is clearly demonstrated in the results from short-term in vitro studies (See Table 4-1).

Moreover, data are needed about

- The chemical nature of the direct-acting mutagens in the diesel exhaust particulate extracts and their potency relationships in bacterial and mammalian cells;
- The bioavailability of mutagenic and carcinogenic substances from inhaled or ingested diesel exhaust particulates;
- The transport of particulates or "biologically significant" levels of released mutagens to critical sites, such as DNA in somatic or germinal cells. Evidence for their direct interaction with DNA in covalent binding would be most useful; and
- In vitro studies for clastogenicity.

TABLE 4.1 Comparison of Mutagenic Activity of Extracts from Different Sources Standardized for Total Organic Matter at 100 µg of Organic Material[a]

	TA98		TA100	
Sample	+S9	−S9	+S9	−S9
Diesel Engine Exhaust Extracts				
Caterpillar	59.3	65.9	115.2	167.8
Nissan	1,367.1	1,225.2	881.7	1,270.1
Oldsmobile	318.1	614.8	169.9	247.5
Volkswagen Rabbit	297.5	399.2	426.0	641.6
Gasoline Engine Exhaust Extracts				
Mustang II	341.9	137.8	228.0	196.5
Comparative Samples				
Cigarette-smoke condensate	98.2	Neg.	--	Neg.
Coke-oven emissions	251.6	164.1	265.6	259.4
Roofing tar	98.7	Neg.	420.0	Neg.
Control Compound				
Benzo[a]pyrene	15,202.3[b]	NT[c]	26,438.0[a]	NT[c]

[a] A linear regression line was developed from the linear portions of the dose-response curves for positive test samples. The equation of that line was used to calcuate the expected response to specific activity at 100 µg of organic material.
[b] Extrapolation.
[c] Not tested.

Source: Claxton (1979).

The following conclusions can be made from short-term and _in vivo_ genetic studies:

• Mutagenic compounds, both direct-acting and S9-activated mutagens, adhere to the central carbonaceous core of diesel exhaust particulates.
• Studies involving _in vitro_ mammalian cell systems indicate that particulates in the exhaust of some diesel engines may contain sufficient levels of biologically available carcinogens to produce cell transformation under conditions of high exposure.
• Biologically active amounts of these substances may be released from particulates that are inhaled or ingested. However, based on available evidence, whole diesel exhaust,

like the whole emissions from comparable gasoline engines, does not appear to be mutagenic in mammals.

- In limited comparisons with extracts from exhaust particulates collected from gasoline engines, diesel exhaust particulate extracts appear to contain more direct-acting bacterial mutagens. However, the activity relationships may not hold in the case of mammalian cell mutation.

CARCINOGENESIS

An important question is whether air pollutants emitted by diesel engines increase the risk of developing lung cancer in humans.

The respiratory tract need not necessarily be the only target organ for inhaled particulates. Following deposition in the air passages of the lungs, the particulates (or materials associated with them) can reach other parts of the body by way of the bloodstream and the lymphatics, as well as by mucociliary clearance to the digestive tract. Thus, lung cancer is not the only potential health hazard associated with exposure to diesel engine emissions. It is, however, the one of most frequent concern.

The question at the start of this section has two parts: Does diesel engine exhaust contribute substantially more carcinogenic material to the environment than gasoline engine exhaust, and does diesel engine exhaust act synergistically with existing carcinogenic agents to which humans are exposed? This, in turn leads to additional questions that studies relating to the carcinogenicity of diesel exhaust need to address:

- Does diesel engine exhaust contain carcinogenic or cocarcinogenic substances--i.e., do recent studies support previous observations that diesel exhaust contains carcinogens?
- What is the chemical nature of the major carcinogens, tumor initiators, and cocarcinogens in diesel exhaust?
- What is the relative carcinogenicity of diesel engine exhaust compared with gasoline engine exhaust?
- What are the most important factors controlling the formation of carcinogens in diesel engine exhaust?
- Are diesel engine exhaust materials carcinogenic for respiratory tract tissues and other organs?

Past and current studies either do not address some of the important matters or do not provide sufficient information. Thus:

- Essentially no information is available concerning the carcinogenicity of gas-phase components. Attempts should be made to learn more about these substances.
- Further identification is needed for the components of diesel exhaust fractions that contain mutagenic and carcinogenic activity. The chemical characterization of these materials would be useful to guide future attempts at engine modification to reduce carcinogenic emissions.

- One of the major questions that has not been adequately resolved concerns the in vivo bioavailability of the toxic substances adsorbed on diesel exhaust particulates. In general, available information seems to indicate that the organic substances are tightly bound to the carbonaceous core. Although they are extractable with polar solvents, such as methylene chloride, the evidence from several experiments suggest that these materials are not bioavailable. In vitro studies, including one with xeroderma pigmentosum cells, as well as two inhalation studies, suggest that some of the materials associated with diesel exhaust particulates are bioavailable. Because of the contradictory findings, studies should be designed to measure, for example, the elution of polycyclic aromatic hydrocarbons from the diesel exhaust particulates in vivo.
- Future investigations need to place more emphasis on comparative studies of the relative carcinogenicity of light-duty diesel engine exhaust and the exhausts of comparable gasoline engines (with and without catalytic converters). This is essential because the decisive issue of whether diesel engines add more potentially carcinogenic agents to the environment per mile driven than gasoline engines under the same load needs to be resolved.
- At present, the most quantitative carcinogenesis data can be expected to be derived from the skin carcinogenesis and the intraperitoneal injection studies with Strain-A mice. Future studies should also make use of another highly sensitive bioassay model--i.e., the newborn mouse (Asahina et al., 1972).

In summarizing the research findings of the current experimental studies related to the potential carcinogenicity of diesel engine exhausts, it must be emphasized that much of the recent work is still incomplete. Thus, final conclusions cannot be drawn as yet. Some of the most important in vivo carcinogenesis studies are currently in progress. Nevertheless, based on the available data, some definitive and some tentative conclusions can be drawn:

- Extracts from diesel (and from gasoline engine) exhaust particulates contain carcinogenic materials. This is supported by many older as well as current chemical and biological studies. The carcinogenic activities of these extracts appear to be two or three orders of magnitude lower on a weight-to-weight basis than that of benzo(a)pyrene, a representative carcinogenic polycyclic aromatic hydrocarbon.
- Whether whole engine exhaust particulates (from gasoline and diesel engines) are carcinogenic is not yet known. Existing data are limited and are either negative or ambiguous. Important studies are under way, involving intraperitoneal injection into Strain A mice and intratracheal injection into Syrian golden hamsters.

- Neither diesel nor whole gasoline engine exhaust has so far been found to be carcinogenic when inhaled by laboratory animals. This negative finding is based mostly on previous studies with a variety of animal species (mice, rats, hamsters, and dogs). Chronic large-scale inhalation studies that are presently under way have not, as yet, yielded information concerning carcinogenicity.

- Variations in fuel composition and engine operating characteristics may turn out to be significant determining factors in the biological activity of diesel exhaust substances according to in vitro cell transformation assays. Whether results from ongoing in vivo studies of diesel particulate matter for carcinogencity will show this phenomenon await the completion of the tests.

- Based on the skin carcinogenesis studies of Misfeld and Timm (1978), Brune and coworkers (1978), and Misfeld (1979), in which the carcinogenic activity of diesel and gasoline engine exhaust extracts (the gasoline engine used was not equipped with a catalytic converter) were compared, it appears that per mile traveled (or on a weight-to-weight basis), the amount of carcinogenic material emitted might be within the same order of magnitude for both types of engines.

- Based on the available data from EPA-supported skin tumor initation studies on SENCAR mice (Slaga et al., 1979), the biological activities of extracts of roofing tar, of coke-oven effluent, and of the exhaust materials from one gasoline engine and two diesel engines are all within the same order of magnitude per unit weight of material tested (See Table 4-2). This comparison does not take into consideration the environmental concentrations of the various effluents to which humans are actually exposed.

- Mouse skin carcinogenicity data and other quantitative bioassay data can be used to estimate the relative carcinogenicity of organic extracts of both diesel exhaust and related environmental emissions. The estimates can then be combined with available data on disease and death patterns in attempting to determine the potential human cancer risk from exposure to diesel engine emissions. Harris (1981) has performed such an epidemiologic assessment. It is based on epidemiologic data of occupational exposure to coke-oven and roofing-tar emissions, along with the results of initiation-promotion experiments on mouse skin and oncogenic transformation experiments from ongoing EPA studies. The resulting estimates of the potential range of lung cancer risk are of the same order of magnitude as those obtained from an epidemiologic study of lung cancer among workers in diesel bus garages in London, England (Waller, 1979). This method of comparative risk assessment assumes that the relative potencies of environmental emissions are preserved across human and nonhuman biological systems. Such an assumption, however, is based on many unknowns. The practical value of

TABLE 4.2 Comparison of Mouse Skin Tumor-Initiating Activities of
Extracts of Diesel Exhaust and Related Environmental Emissions[a]

Sample	Papillomas/mouse/mg[b] (14 weeks)	R^2
Caterpillar diesel engine	0	--
Nissan diesel engine	0.258	0.996
Oldsmobile diesel engine	0.115	0.95
Mustang II gasoline engine	0.09	--
Cigarette-smoke condensate	0	--
Roofing tar	0.182	0.999
Coke-oven emissions	0.307	0.876
Benzo[a]pyrene	46.2	0.984

[a] In the tests, 40 males and 40 females of SENCAR mice were initiated
with the various samples and promoted by twice-weekly applications of
2 μg tetradecanoyl phorbol acetate.

[b] The values represent the slope from the linear regression analysis of
the dose-response studies and the correlated "measure of fit" (R^2).

Source: Slaga et al. (1979).

risk assessments relying on such assumptions is limited in
view of interspecies and interorgan differences in factors
such as bioavailability, particulate distribution,
extractability and clearance of active organics, target site
of action, metabolism, and genetic repair mechanisms.

Despite the shortcomings of the assay systems used and the
incompleteness of the data, three findings emerge from the various
chemical and biological studies performed on diesel emissions. They
are: diesel exhaust contains traces of carcinogenic materials; the
carcinogenic activity of these materials appears to be low; and
variations in engine operating characteristics and fuel type appear to
affect the carcinogenic activity of diesel exhaust particulates.

PULMONARY AND SYSTEMIC EFFECTS

Evaluating the potential pulmonary and systemic effects of exposure
to diesel exhaust should be done with emphasis on anticipated morbidity
rather than mortality. In this section the Health Effects Panel
examines past, ongoing, and proposed research on the relationship of
various gaseous and particulate components in diesel exhaust with
pulmonary and systemic problems in humans. Following are the major
concerns:

- Diesel exhaust particulates are small enough (< 1 μm) to be readily deposited deep in the lungs. The particulates thus have the potential for carrying toxic substances into the lungs where they may be leached off and transported by way of the systemic circulation into other organs.
- The extent to which the various materials are soluble in body fluids--that is, their bioavailability and its relationship to their ultimate systemic toxicity;
- The extent to which the gaseous portion of the exhaust could adversely alter the ambient levels of certain pollutants (CO, SO_x, NO_x, O_3, and volatile aldehydes), depending on the nature of the fuel and the condition of the engine; in addition, the extent to which components of the gaseous portion would undergo atmospheric chemical reactions (as in photochemical smog) that create other toxic volatile substances (e.g., peroxyacetyl nitrate);
- The effects of diesel exhaust on specific populations such as those with cardiopulmonary diseases, asthmatics, the very young, the old; and the effects of differences in levels of exposure to both the gaseous and the particulate components under varying conditions of physical activity such as recreational exercise and vigorous work;
- The noxious effects of the total exhaust, especially odor and visibility, and possibly eye irritation, on various exposed populations;
- The potential for an increase of cardiovascular diseases resulting from the possible incremental addition of CO to present ambient levels;
- The potential for a rise in infectious diseases in the very young (< 2 years) by the additional NO_2 in the ambient air; and
- The suggested evidence for adverse behavioral effects caused by the influence of certain of the gaseous components on the human central nervous system.

The data on the pulmonary and systemic health effects of exposure to diesel exhaust are extremely limited. The following are the most obvious research gaps:

- Information is lacking on the acute toxicity of diesel exhaust. A reevaluation of the acute effects of diesel exhaust on lungs should emphasize exposure to exhausts with different characteristics that are generated by varying the modes of engine operation and by using different fuels. Primary lung damage and recovery should be fully documented with quantitative morphologic techniques and selected physiologic and biochemical studies (airway resistance, induction of protective enzymes, etc.). None of the presently conceived studies has considered the usefulness and value of detailed cell kinetic studies. These are of particular assistance in quantitating initial cell death in the lung (Evans et al., 1978).

- It is necessary to determine possible long-term consequences of diesel exhaust inhalation, such as the development of fibrotic and emphysematous changes in the lung. Integral to this is the knowledge of whether lesions are reversible upon the cessation of exposure. To provide this information, different animal species should be exposed to graded concentrations of diesel exhaust in studies of several months duration and the extent and degree of induced changes should be fully documented.

- Additional quantitative data need to be obtained on initial deposition and clearance of inhaled particulates, possible translocation to other organs and tissues, and retention in the body. The respiratory tract need not necessarily be the only organ affected by inhaled particulates from diesels. Particulates (and material associated with them) can reach other parts of the body through the bloodstream and the lymphatics, as well as by mucociliary clearance to the digestive tract. The leaching of potentially toxic compounds from the particulates needs to be determined and related to potential systemic effects. A single study reporting that inhaled diesel exhaust causes biochemical changes in extrapulmonary tissue (Lee et al., 1980) suggests the need for additional studies.

- Increased susceptibility of animals to infection following inhalation of diesel exhaust must be evaluated in young, mature, and old animals. An additional issue is whether resistance to a bacterial or viral challenge is modified primarily by the gaseous phase or by the particulate fraction of diesel exhaust. The functional biology of particulate-laden macrophages, and the overall capacity of the macrophage system to handle and remove particulate material under conditions of continuous exposure, needs to be thoroughly investigated. This includes cell kinetic studies on the biology of macrophages (Adamson and Bowden, 1980) and quantitative morphometric studies. The effects of diesel exhaust on the immune system, on the lung, and on other organs should be evaluated with appropriate techniques (Vos, 1977).

- It is important to evaluate how diesel exhaust affects humans with preexisting diseases. For instance, the presence of pulmonary emphysema has been shown in one study to alter deposition and long-term clearance of inhaled particulates in hamsters (Hahn and Hobbs, 1979). That no experiments are planned with animals suffering from conditions similar to certain human diseases is clearly a research gap. Such animal models exist--e.g., pulmonary emphysema (Karlinsky and Snider, 1978), pulmonary fibrosis (Snider et al., 1978; Haschek and Witschi, 1979), immunosuppression from cigarette smoking (Holt et al., 1978), alveolar lipoproteinosis (Heppleston, 1975), chronic pulmonary hypertension (Kentera et al., 1978), increased sensitivity to ozone (Calabrese, 1978), and cardiomyopathy with general heart failure (Gertz, 1973). The

need is urgent to develop and use animal models of human
diseases in order to relate the effects of specific primary
and secondary products from diesel exhaust to specific
disorders in humans.

Based on available information, few conclusions regarding pulmonary
and systemic effects can be drawn. This follows from the paucity of
information about the effects of diesel emissions on human health, as
well as the preliminary state of the experimental work in progress.

- The acute and chronic inhalation of diesel exhaust produces,
  as expected, the accumulation in the deep lung of carbonaceous
  particulates, as well as potentially hazardous compounds
  adsorbed to them. Such materials become sequestered primarily
  in alveolar macrophages and, to a limited extent, in cells of
  the alveolar epithelium. Clearance may occur via the
  mucociliary escalator and the pulmonary lymphatic system. The
  possible long-term consequences of such accumulation with
  regard to its potential for causing chronic pulmonary disease
  is a key issue in the evaluation of diesel exhaust inhalation
  hazards. Furthermore, there is the question of whether
  adverse health effects may be exacerbated if synergistic
  interactions occurring in the environment--e.g., those between
  diesel exhaust particulates and products of photochemical
  reactions--increase the toxicity of exhaust compounds.
  Experimental data are insufficient to resolve this question.
- Histopathologic changes elicited by inhaled diesel exhaust are
  nonspecific. They may be interpreted to reflect initial cell
  damage followed by recovery with discrete areas of fibrosis
  and possibly emphysematous changes. The current data confirm
  that the fibrogenic potential of diesel exhaust is low.
  However, additional lifespan exposure data are needed.
- A single observation suggests that inhaled material may induce
  biochemical changes in organs distant from the respiratory
  tract (Misiorowski et al., 1980). Because these materials are
  cleared relatively slowly, studies following inhalation
  exposure need to be of sufficient duration to determine the
  secondary effects of inhaled materials.
- Present information suggests that pulmonary defense mechanisms
  may be adversely affected by diesel exhaust. It is not clear
  whether the agents responsible for this phenomenon are
  associated with the gaseous phase or the particulate phase of
  the exhaust. Low levels of NO_x exposure have been shown to
  decrease resistance to infectious diseases in both animals and
  humans.
- Available information suggests that a single high-level
  exposure to diesel exhaust can produce acute toxic
  effects--e.g, poisoning due to NO_x, to aldehydes, and
  possibly to CO--whereas long-term exposure to comparatively
  low diesel exhaust levels has not clearly been shown to cause
  pulmonary and systemic toxicity. The determination of

ultimate health effects requires consideration of the data bases on studies involving both acute and chronic studies

- Analysis of the available experimental evidence of pulmonary and systemic health effects caused by exposure to diesel exhaust suggests that it is possible to estimate the health hazards from the expected increase in gaseous and particulate components in the general atmosphere. With respect to pulmonary and systemic effects, it is reasonable to expect that health hazards associated with certain pollutants originating in diesel exhaust (SO_x, NO_x, O_3, CO, and possibly particulate material) would be qualitatively similar to those associated with the same pollutants from other sources--e.g, fossil-fueled power plants.

EPIDEMIOLOGY

Epidemiology is concerned with relationships between environmental exposures and disease frequency and distribution patterns in clearly defined groups of people. The principal question is: What adverse effects can be anticipated from the increasing use of diesel-powered, light-duty vehicles on the health of the U.S. population during the next two decades?

One way to answer the question is to monitor exposure levels and trends for diesel exhaust in different places and relate these to the levels and trends in the health of the public. Studies of this kind are difficult to carry out successfully. It is difficult, for example, to anticipate all the other relevant changes that may occur over the next decade or more and to allow adequately for them. Moreover, unless it proves possible to identify a population living in an area where nearly a quarter of the light-duty vehicles are diesel-powered, for comparison with one in which the proportion is much lower, no conclusion is likely to be obtained for 10 to 20 years, if then.

An alternative approach is to focus on population groups that have been exposed at work to high concentrations of diesel exhaust for long periods. The health of workers employed in bus garages, rail transport facilities, and underground mines where diesel engines are used can be compared with the health of workers at similar socioeconomic levels in jobs that require comparable physical effort but do not involve exposure to diesel exhaust. If an adverse effect is found, it may be possible, by studying exposure-response relationships, to estimate the magnitude of effects on the working population. By making certain assumptions, it is also possible to estimate what the effects might be for the general population. One important proviso that should be considered is that there may be persons in the general population, but not in the working population, who are exceptionally susceptible to diesel exhaust emissions and are therefore at greater risk than the general population.

Two critical questions about diesel exhaust need to be addressed in epidemiological studies:

- Does it cause cancer -- more specifically, lung cancer?
- Does it cause chronic, nonmalignant respiratory disease?

If the answer to either of these questions is yes, the risk must be quantified by establishing exposure-response relationships. The following conclusions can be appropriately drawn from the review of the literature on diesel exhaust exposures:

- In epidemiologic studies of occupational exposure to diesel engine emissions, excess risk of cancer of the lung, or of any other site, has not been convincingly demonstrated. The evidence to date does not indicate that exposure to diesel exhaust is a serious cancer hazard, at least at exposure levels no greater than those that existed in London bus garages (Waller, 1979). Only two studies, one on railroad workers (Kaplan, 1959) and the other on bus garage workers (Raffle, 1957; Waller, 1979), approximate even the minimum requirements for a sound epidemiologic evaluation of cancer risk. Both suffer from deficiencies in design. Hence, their negative conclusions should be viewed with caution.
- Evidence of a relation between occupational exposure to diesel exhaust and prevalence of chronic obstructive lung disease is inconsistent. Some studies have suggested that workers exposed to diesel exhaust have a higher prevalence of chronic respiratory symptoms and bronchitis, and diminished lung function than otherwise comparable persons who have not been exposed. Other studies have failed to confirm these observations. Because of this inconsistency in the findings, it remains uncertain whether exposure to diesel exhaust was a very important factor in the development or exacerbation of chronic respiratory disease in the population groups studied.
- Additional carefully controlled studies of populations occupationally exposed to diesel engine exhaust are needed. In such studies, both the whole exhaust and its individual components should be carefully monitored. The studies need to be carefully controlled for cigarette smoking, which plays a dominant role in the etiologies of lung cancer and chronic obstructive lung disease.
- Several epidemiological studies have suggested a synergism between cigarette smoking and occupational exposure in the development of lung cancer (International Union Against Cancer, 1976). Asbestos workers and uranium miners who smoke appear to be exceptionally prone to develop this cancer (Selikoff et al., 1968, 1980; Archer et al., 1973). Synergism between smoking and diesel exhaust might similarly increase the risk of lung cancer, although this has not yet been shown to occur. Future epidemiologic researchers should keep this possibility in mind.

REFERENCES

Adamson, I., and D. Bowden (1980). Role of Monocytes and Interstitial Cells in the Generation of Alveolar Macrophages. II. Kinetic Studies after Carbon Loading. Laboratory Investigation. 42:515-524.

Ames, B. N., W. E. Durston, E. Yamasaki, and F. D. Lee (1973). Carcinogens are Mutagens: A Simple Test System Combining Liver Homogenates for Activation and Bacteria for Detection. Proceedings of the National Academy of Sciences. 70:2281-2285.

Ames, B. N. (1979). Identifying Environmental Chemicals Causing Mutagens and Cancer. Science. 204:587-593.

Archer, Z., J. Wagonner, and S. London, Jr. (1973). Uranium Mining and Cigarette Smoking Effects on Man. Journal of Occupational Medicine. 15:204-211.

Asahina, J., A. Carmetl, E. Arnold, Y. Bishop, S. Joshi, D. Coffin, and S. Epstein (1972). Carcinogenicity of Organic Fractions of Particulate Pollutants Collected in New York City and Administered Subcutaneously to Infant Mice. Cancer Research. 32:2263-2268.

Bouch, N., and G. di Mayorca (1976). Somatic Mutation as the Basis for Malignant Transformation of BHK Cells by Chemical Carcinogens. Nature. 264:722-727.

Brune, H., M. Habs, and D. Schmal (1978). The Tumor-Producing Effect of Automobile Exhaust Condensate and Fractions Thereof. Part II. Animal Studies. Journal of Environmental Pathology and Toxicology. 1:737-746.

Brusick, D. (1978). The Role of Short-Term Testing in Carcinogen Detection. Chemosphere. 7(5):413-417.

Calabrese, E. (1978). Animal Model of Human Disease. Increased Sensitivity to Ozone. Animal Model: Mice with Low Levels of G-6-PD. American Journal of Pathology. 91:409-411.

Claxton, L. (1980). Mutagenic and Carcinogenic Potency of Diesel and Related Environmental Emissions: Salmonella Bioassay. Health Effects of Diesel Engine Emissions: Proceedings of an International Symposium. Volume 2. EPA-600/9-80-057b.

Evans, M., N. Dekker, L. Cabral-Anderson, and G. Freeman (1978). Quantitation of Damage to the Alveolar Epithelium by Means of Type 2 Cell Proliferation. American Review of Respiratory Disease. 118:787-790.

Gertz, E. (1973). Animal Model of Human Disease. Myocardial Failure, Muscular Dystrophy. Animal Model: Cardiomyopathic Syrian Hamster. American Journal of Pathology. 70:151-154.

Hahn, F., and C. Hobbs (1979). The Effect of Enzyme-Induced Pulmonary Emphysema in Syrian Hamsters on the Deposition and Long-Term Retention of Inhaled Particles. Archives of Environmental Health. 34:203-210.

Hare, C., and T. Baines (1979). Characterization of Particulate and Gaseous Emissions from Two Diesel Automobiles as Functions of Fuel and Driving Cycle. SAE Technical Paper 790424.

Harris, J. (1981). Potential Risk of Lung Cancer from Diesel Engine Emissions. Report prepared for the Diesel Impacts Study Committee,

79

National Research Council. Washington, D.C.: National Academy Press.

Haschek, W., and H. Witschi (1979). Pulmonary Fibrosis - A Possible Mechanism. Toxicology and Applied Pharmacology. 51:475-487.

Heppleston, A. (1975). Animal Model of Human Disease. Pulmonary Alveolar Lipo-Porteinosis. Animal Model: Silica-Induced Pulmonary Alveolar Lipo-Proteinosis. American Journal of Pathology. 78:171-174.

Holt, P., D. Keast, and J. MacKenzie (1978). Animal Model of Human Disease. Infectious and Neoplastic Respiratory Diseases Associated with Cigarette Smoking. Animal Model: Immunosuppression in the Mouse Induced by Long-Term Exposure to Cigarette Smoke. American Journal of Pathology. 90:281-284.

Huisingh, J., R. Bradow, R. Jungers, L. Claxton, R. Zweidinger, S. Tejada, J. Bumgarner, F. Duffield, M. Waters, V. Simmon, C. Hare, C. Rodriquez, and L. Snow (1978). Application of Short-Term Bioassays in the Fractionation and Analysis of Complex Environmental Mixtures. M. Waters, S. Nesnow, J. Huisingh, S. Sandu, and E. Claxton (eds.). Research Triangle park, N.C.: EPA, Health Effects Research Laboratory. EPA-600/9-78-027.

International Union Against Cancer (1976). Animal Models. Lung Cancer. A Series of Workshops on the Biology of Human Cancer. Report #3. Geneva: UICC Technical Report Series. 25:1-30.

Jorgensen, H., and A. Svensson (1970). Studies on Pulmonary Function and Respiratory Tract Symptoms of Workers in an Iron Ore Mine Where Diesel Trucks are Used Underground. Journal of Occupational Medicine. 12:349-354.

Kaplan, I. (1959). Relationship of Noxious Gases to Carcinoma of the Lung in Railroad Workers. Journal of the American Medical Association. 171:2039-2043.

Karlinsky, J., and G. Snider (1978). Animal Models of Emphysema. American Review of Respiratory Disease. 117:1109-1133.

Kentera, D., D. Susic, M. Zdravkovic, V. Kanjuh, and G. Tucakovic (1978). Chronic Pulmonary Hyptertension. Comparative Pathology Bulletin. 10: 2, 4.

Kotin, P., H. Falk, P. Mader, and M. Thomas (1954a). Aromatic Hydrocarbons. I. Preserve in a Los Angeles Atmosphere and the Carcinogenicity of Atmosphere Extracts. Archives of Industrial Health. 9:153-163.

Kotin, P., H. Falk, and M. Thomas (1954b). Aromatic Hydrocarbons. II. Presence in the Particulate Phase of Gasoline-Engine Exhausts and the Carcinogenicity of Exhaust Extracts. AMA Archives of Industrial Health. 9:164-177.

Lee, I., K. Suzuki, S. Lee, and R. Dixon (1980). Aryl Hydrocarbon Hydroxylase Induction in Rat Lung, Liver and Male Reproductive Organs Following Inhalation Exposure to Diesel Emission. Toxicology and Applied Pharmacology. 52:181-184.

McCann, J., E. Choi, E. Yamasaki, and B. Ames (1975). Detection of Carcinogens as Mutagens in the Salmonella/Microsome Test: Assay of 300 Chemicals. Proceedings of the National Academy of Sciences. 72:5135-5139.

Magee, P. (1977). The Relationship Between Mutagenesis, Carcinogenesis and Teratogenesis. Progress in Genetic Toxicology. D. Scott, B. Bridges, and F. Sobels (eds.). Amsterdam: Elsevier/North Holland Biomedical Press. pp. 15-27.

Miller, E., and J. Miller (1971). The Mutagenicity of Chemical Carcinogens: Correlations, Problems, and Interpretations. Chemical Mutagens: Principals and Methods for Their Detection. Vol. 1. A. Hollaender (ed.). New York: Plenum Press. pp. 83-119.

Misfield, J. (1980). The Tumor Producing Effects of Automobile Exhaust Condensate and of Diesel Exhaust Condensate. Mathematical-Statistical Evaluation of the Test Results. Health Effects of Diesel Engine Emissions: Proceedings of an International Symposium. Volume 2. EPA-600/9-80-057b.

Misiorowski, R. L., K. A. Strom, J. J. Vostal, and M. Chvapil (1980). Lung Biocehmstry of Rats Chemically Exposed to Diesel Particles. Health Effects of Diesel Engine Emissions: Proceedings of an International Symposium. Volume 1. EPA-600/9-80-057a.

Raffle, P. (1957). The Health of the Worker. British Journal of Industrial Medicine. 14:73-80.

Selikoff, I., E. Hammond, and J. Chung (1968). Asbestos Exposure, Smoking and Neoplasia. Journal of the American Medical Association. 204:106-112.

Selikoff, I., H. Seidman, and E. Hammond (1980). Mortality Effects of Cigarette Smoking Among Amocite Asbestos Workers. Journal of the National Cancer Institute. 65:507-513.

Slage, T., L. Triplett, and S. Nesnow (1980). Mutagenic and Carcinogenic Potency of Extracts of Diesel and Related Environmental Emissions: Two-Stage Carcinogenesis in Skin Tumor Sensitive Mice (Sencar). Health Effects of Diesel Engine Emissions: Proceedings of an International Symposium. Volume 2. EPA-600/9-80-057b.

Snider, G., J. Hayes, and A. Korthy (1978). Chronic Interstitial Pulmonary Fibrosis Produced in Hamsters by Endotracheal Bleomycin. American Review of Respiratory Disease. 117:1099-1108.

Vos, J. (1977). Immune Suppression as Related to Toxicology. CRC Critical Reviews in Toxicology. 5:67-101.

Waller, R. (1980). Trends in Lung Cancer in London in Relation to Exposure to Diesel Fumes. Health Effects of Diesel Engine Emissions: Proceedings of an International Symposium. Volume 2. EPA-600/9-80-057b.

Williams, R., and S. Swarin (1979). Benzo[a]pyrene Emissions from Gasoline and Diesel Automobiles. Environmental Science and Analytic Chemistry Departments. Warren, MI: General Motors Research Laboratories. GMR-2881R.

5 ECONOMIC EFFECTS

From the user's perspective, the most attractive feature of diesel cars is that their fuel economy surpasses that of gasoline-burning cars of similar size and horsepower. Still, the fuel economy of diesels varies with engine type and with road and driving conditions. Diesels also tend to have less power and acceleration than gasoline-powered vehicles of comparable engine displacement or size. Hence, it is difficult to make precise statements about the fuel economy advantages of diesels.

Compared with gasoline engines with similar performance characteristics, diesels offer a fuel economy advantage in mpg of 25 to 35 percent, and compared with gasoline engines of similar displacement or size, diesels provide a fuel economy advantage of 30 to 40 percent. Table 2.1 indicates some relevant comparisons. The fuel economy comparisons are based on fuel consumption measurements made by EPA. In actual road use, however, the fuel economy of gasoline-powered vehicles is somewhat less than reported in the EPA tests--perhaps by as much as 15 percent (McNutt et al., 1979). By contrast, diesels do not seem to encounter the same fuel economy differences between the EPA tests and actual use.

Thus, diesel fuel economy provides monetary savings for owners of diesel cars. Furthermore, diesels may be attractive to buyers who fear that oil and its products may be in limited supply in the future. Some owners may believe that diesel fuel, which is used also in trucks and agricultural tractors, may be less tightly rationed than gasoline in the event of oil shortages.

Another advantage of diesels is that, historically, diesel engines for trucks have tended to be more durable than gasoline engines. Whether this will hold true for passenger cars and small trucks is not known.

Motorists also recognize some unfavorable sides to diesel vehicles. Diesels are more noisy and less powerful than gasoline engines of comparable size. They are more difficult to start in cold weather. Their exhaust systems frequently emit noticeable clouds of smoke, which usually have a perceptible (and, to many, unpleasant) odor. Finally, some diesels currently require more frequent and more expensive maintenance than vehicles with spark-ignition engines.

These qualities apply to diesels in current production and use. In the United States, the number of diesels in today's automotive fleet is small—somewhat less than 1 million of the 108 million cars on the roads. If the market demand for diesel passenger cars increases substantially, the financial conditions of the manufacturers improves, and the regulatory uncertainties confronting the car makers are resolved, it is most likely that diesel research and development efforts will increase and that, as a result, most of the diesel's present technological difficulties will be overcome. Among those, the cold-starting and the exhaust smoke problems are likely to be corrected, even without the use of particulate control devices. In addition, further improvements are possible in fuel economy—both absolute and in comparison with gasoline engines.

COSTS

Today's diesel-powered passenger cars and light trucks are priced higher than comparable gasoline-powered vehicles. The price difference is in the range of $350 to $800 per vehicle. How much of this added price reflects added manufacturing costs is unclear. There are costs of new mechanical components and extra manufacturing efforts—e.g., the fuel injection system, heavier engine block and head, stronger pistons and crankshaft, and a more powerful battery. There are savings in unnecessary devices—notably the carburetor, spark plugs and related electrical system components, and catalytic converter. Recently, the added price to consumers has exceeded added costs. Accordingly, in 1979 and 1980 diesel cars sold at a premium. Their relative scarcity was caused by an unexpected surge in demand following the sharp increase in gasoline prices in mid-1979. As manufacturers increase their outputs of diesels, the premiums should disappear, and price differences approximating the true costs should appear. The price disparity might fall in the range of $300 to $600 for a diesel car over a similar sized vehicle with a conventional gasoline engine.*

Most diesels, especially those with larger engines, appear to have higher maintenance costs than comparable gasoline-powered vehicles. For instance, there are no spark plugs and carburetors to replace in diesels, but diesels usually require more frequent oil changes and their crankcases hold more oil. The added maintenance costs over a 100,000-mile lifetime of a diesel appear to range from a negligible amount for the smallest diesels to $400 to $500 (undiscounted) for the largest.

Table 5.1 shows a range of savings over the expected 100,000-mile life of three different sizes of diesel vehicle. The first source of

---

*If the market for diesel cars increases, economies of scale and manufacturing improvements are likely to come about, so that the price differential between diesel-fueled and gasoline-fueled vehicles would disappear and lifetime cost savings for diesel owners would increase. The anticipated savings would probably lead to greater demand for diesel cars—though wider use of diesels would not substantially affect the results of the analysis of alternative diesel particulate emission standards presented in Chapter 7.

TABLE 5.1   Net Present Value of Fuel Cost Savings per Diesel Cars
(10% Discount Rate, 10-year or 100,000-Mile Vehicle Life)

| | Size of Vehicle | | |
	Large	Medium (Compact)	Small (Sub-compact)
Diesel vehicle			
Fuel efficiency (mpg)	25-28	35-39	45-50
Fuel use (U.S. gallons)	4,000-3,500	2,850-2,550	2,200-2,000
Gasoline vehicle			
Fuel efficiency (mpg)	20	28	36
Fuel use (U.S. gallons)	5,000	3,550	2,800
Fuel cost savings (dollars)			
Prices of diesel fuel and gasoline equal ($1.25 per gallon in 1980)*	1,000-1,500	750-1,050	550-800
Additional dollar savings if price of diesel fuel $0.10 per gallon less than price of gasoline	330-330	240-210	180-160
Fuel-related dollar saving (including fuel price differential)	1,330-1,800	990-1,260	730-960

*Fuel costs are presumed to rise in real terms at a rate of 5 percent per year from $1.25 a gallon in 1980 to $2.00 a gallon in 1990.

savings is greater fuel economy. A range of 25 to 40 percent is used as the likely increase in miles per gallon for diesels, and a representative 1980 price of $1.25 for a gallon is used to estimate the cost of gasoline. The second source of saving comes from the lower purchase price of diesel fuel. In most communities, diesel fuel sold for 5 to 15 cents less per gallon in 1980 than unleaded regular gasoline; the average saving was about 10 cents. It is possible that this difference will disappear in the near future (see Chapter 2).

The lifetime cost savings shown in Table 5.1 are based on the assumption that the price of gasoline rises from $1.25 a gallon in 1980 at a rate of about 5 percent per year (in constant 1980 dollars) to about $2.00 in 1990, when cars bought in 1980 will be retired, on average, from service. Table 5.1 shows the net present value of savings in fuel outlays in constant dollars for a 10 percent discount rate. This rate may be taken as the "real" rate (net of inflation) reflected in individual investment and consumption choices, which may govern buyer behavior in the current diesel market. A lower rate, such as 5 percent, would constitute the "real" social discount rate that ought to be used in making public policy choices. The difference is a result, in part, of the existence of taxes on individuals that are merely transfers and not real costs from the point of view of society as a whole.*

Table 5.1 also shows the effect on lifetime cost savings if there is a price difference between gasoline and diesel fuel in relative terms over the lifetime of vehicles purchased in 1980. The lifetime fuel cost differences are shown for alternative assumptions about the fuel efficiency advantage of diesels and for variations in the size classes of vehicles.

The lifetime fuel cost savings, although significant for all size classes, is quite sensitive to vehicle size. It arises primarily from the fuel efficiency advantage of diesels, rather than from the lower price of diesel fuel. Thus, whatever the uncertainty about the persistence of the present cost advantage of diesel fuel, the fuel economy of diesels promises significant savings in fuel costs over the lifetime of the vehicle.

In the life of a typical large diesel car, then, fuel savings could amount to as much as $1,800. For the smallest diesel car, the savings may be around $950. The estimates are conditional on car size and operation for 100,000 miles. Unconditionally, diesel cars may last even longer and the price difference of diesel fuel and gasoline may disappear.

CONSEQUENCES FOR SAFETY

For passenger cars and other motor vehicles, any reduction in inertia weight improves the fuel economy. (Such relationships have been discussed in Chapter 2.) In fact, most of the fuel economy gains for automobiles for model years 1976 to 1981 have resulted from reductions in the weight of the car rather than improvements to the power train. Despite the better gas mileage, lighter motor cars have an unfortunate side. Accident statistics indicate that smaller, lighter vehicles expose drivers and passengers to greater risk of serious injury or death in traffic accidents. In a study for the National Highway Traffic Safety Administration in the mid-1970's, Stewart and Stutts (1978) found that occupants of the lightest cars were at almost exactly twice the risk of those in the heaviest automobiles.

Diesels may alter that equation. For a given average fuel economy level for the fleet, the number of traffic injuries and fatalities may decrease with the proliferation of diesel cars, because diesel-powered vehicles can be heavier than gasoline-powered vehicles that provide

---

*Such a treatment of discount rates was adopted by Spurgeon Keeney and his colleagues in Nuclear Power Issues and Choices, A Report of the Nuclear Energy Policy Study Group. Ballinger Publishing Co., Cambridge, Mass., 1977.

similar or better fuel economy. This might result from patterns of driver preference or from binding Corporate Average Fuel Economy (CAFE) standards. Thus, if the diesel powers 25 percent of vehicle miles traveled (VMT) in 1990 (at an average vehicle weight that leaves the fleet fuel economy unchanged by comparison with an all gasoline fleet), we would expect several hundred fewer fatalities and several thousand fewer injuries per year. The actual outcome could well be at an intermediate point, with some gain in fuel economy and some reduction in fatalities. However, it is worth noting that government policy in recent years calls for improving fuel economy through CAFE standards. To the extent that this has influenced car sales, regulatory policy has had the effect of downsizing the fleet and, consequently, accepting an increase in the number of traffic casualties.

To estimate the effect of banning diesels in terms of traffic casualties, the committee adopted a weight-safety model based on three key observations (McDonald and Ingram, 1981):

- The probability of an accident (per VMT) does not depend on vehicle weight, once driver characteristics such as age are considered.
- In multiple-car collisions, heavier vehicles have a decisive advantage when they are involved with cars of equal weight as well as when they encounter smaller cars. Heavier cars have two advantages in collisions: They decelerate less rapidly than smaller vehicles, and they have more interior space to absorb the impact and insulate the occupants. Two independently derived models, based on empirical data, (Mela, 1974; Carlson, 1979), show comparable characterizations of this phenomenon.
- In accidents involving a single vehicle and in collisions of a car and truck, the larger passenger vehicle is apparently safer, but this outcome is less certain than in collisions between multiple vehicles.

If the safety features of passenger cars (improved bumpers, dual braking systems, etc.) are not altered over the likely range of future vehicle sizes, a reduction in average vehicle weight of 100 pounds is estimated to increase the annual number of traffic fatalities by approximately 1,000 (McDonald and Ingram, 1981). In a situation where diesels comprise 10 percent of the light-duty fleet and then are prohibited for reasons of public policy, so that the requirement for fuel economy is made up by weight reductions in conventional gasoline-powered cars, the number of traffic deaths would increase by approximately 800 per year. If the proportion of diesels were to reach 25 percent and diesel cars were to be banned from the roads, the number of traffic fatalities would rise by about 2,000 each year.

Under certain circumstances, the adverse safety effects could be reduced. First, the parameters of the statistical models were estimated from data on accidents involving cars manufactured in the 1970's, and were affected by the safety features in those vehicles, as well as by such occupant habits as failure to use seat belts. In future vehicles, new safety features may mitigate injuries and reduce fatalities by changing the current relationship between traffic casualties and vehicle weight, thus diminishing the effect of downsizing on traffic fatalities. In particular, the uniform front-end crash standards to be implemented in the early 1980's should tend to reduce the safety inequality between large and small cars. It is difficulty to forecast precisely how large an effect such design changes will have, but it is not likely that the present relationship between vehicle weight or size and occupant safety will be altered.

Second, future changes in safety design, such as the introduction of passive restraint systems (e.g., automobile seat belts and air bags), could also reduce the number of road fatalities. Although the effectiveness of future safety design changes is difficult to forecast, a reduction in the level of total fatalities will proportionately reduce the safety effect calculated in the model.

Third, the projections for injuries and deaths assume that the fuel economy of gasoline cars is not affected by a prohibition or limitation of the use of diesels. In fact, technological advantages in gasoline engines might be stimulated if diesels were banned, because more research and development might be applied instead to gasoline engines.

One factor that tends to affect the estimate is the potential for reducing automobile deaths by fire. Fire-related fatalities are difficult to predict because death may have been caused by the collision or by a fire following the crash. Still, the physical properties of diesel fuel tend to retard fire, and in recognition of this insurance discounts are applied to diesel trucks. A review of state data on truck accidents reveals that diesel vehicles have a significantly lower incidence of fiery collisions.

NET BALANCE IN CONSUMER SATISFACTION

Table 5.2 summarizes the estimates of initial costs, maintenance costs, and fuel economy savings for diesels compared with gasoline-powered vehicles over 100,000 miles on the roads. On balance, a diesel owner can expect a modest or substantial net savings, depending on the expected fuel economy advantage and the expected future increase in the real price of fuel. The gains would be greater if the diesel owner expects to drive more than 100,000 miles during the life of the vehicle or places a higher value on the potential reduction of inconvenience from possible fuel rationing schemes or service station lines.

The financial benefits shown in Table 5.2 may overstate the overall benefits to diesel owners, because automobile owners may place greater value on the type of car they buy than on the dollar savings shown in the table. Thus, motorists may decide to sacrifice all or part of the savings attainable from a diesel's fuel economy for the ability to purchase a larger, more comfortable, and safer automobile. For example, a motorist who otherwise might have purchased a medium-sized gasoline-powered car because of the fuel economy advantages over a larger gasoline-powered car might choose a medium-sized diesel-powered car for the net monetary gains shown in Table 5.2. If the additional savings were not considered important, the same motorist might purchase

TABLE 5.2   Present Value of Savings of Owning a Diesel Car Compared
with Gasoline Car (10-year or 100,000-mile Vehicle Life)

| (Costs in U.S. dollars, 1980) | Large | Size of Vehicle | |
		Medium (Compact)	Small (Sub-compact)
Initial cost	-800	-700	-600
Maintenance	-400	-200	0
Fuel	1,330-1,800	990-1,260	730-960
Total Savings	130-600	90-360	130-360

a large-sized diesel-powered car, which would offer at least as good
fuel economy as the medium-sized gasoline-powered car, along with
greater size and comfort. The decision will depend on the value a
buyer places on added fuel economy (at the same vehicle size and
comfort) versus added size, comfort, and safety (at the same fuel
economy). The monetary savings shown in Table 5.2 compare diesel cars
and gasoline cars of similar size. If the availability of diesels
causes some buyers to increase the size of the vehicles they purchase,
the financial savings to those buyers will understate the total benefit
they derive from the availability of diesels. An estimate of the
willingness of consumers to pay for the safety advantages of diesels is
not included because there is no information on the strength of such
preferences--i.e., in this respect, Table 5.2 underestimates consumer
financial benefits. The motorist may choose a larger vehicle instead
of enjoying the full fuel economy advantages of a smaller one--the
choice being dependent on the buyer's values for certain attributes.
Thus, the actual net gains are likely to be somewhat lower.

FUTURE DIESEL SALES AND USE

In the United States, sales of diesel-powered automobiles have
accelerated in the past five years, increasing from 22,421 in 1975 to
387,048 in 1980--or from a fraction of a percent to some 4.6 percent of
all cars sold. If light trucks are included in the car sales figure,
the percentage falls to about 4 percent.
Predicting the extent of the increase in lightweight diesel sales
is difficult because of the uncertainties and unknowns. Future sales
of diesels will be a function of the relative rates of improvements in
fuel economy, manufacturing costs, and operating performance for both
diesel and gasoline engines; the rate of increase in fuel prices; the
extent to which diesel engines are regulated (and the cost of meeting

that regulation); and the extent to which manufacturers are able to convert and construct facilities to build diesels in the 1980's. So, obviously, the number of diesels sold will depend not only on the market share, but on such factors as the relative prices of automobiles and fuels (compared to other things on which consumers might spend their income), and the possible technological changes that might be competitive with automobile travel--e.g., improvements in mass transit or air travel, say, or developments in communications such as cable-TV and home computers that might decrease trips by car to shop, attend movies and sports events, go to the bank, conduct business matters, take part in meetings and conferences.

A range of cases for various levels of diesel sales is presented in Chapter 7, but it is important to state here that levels of sales do not translate directly into levels of vehicles in use. In approximate terms, the automobile fleet turns over about once very 10 years. Hence, a 20 percent level of sales for any particular car will have to be sustained for about 10 years to achieve a 20 percent level of the total number of vehicles in use. Table 5.3 illustrates this situation. If diesel sales were to rise to 20 percent in five years, the percentage in use would be only about 5 percent, and if sales rise more slowly to 20 percent around, say, 1990 then only 8 percent of the vehicles in use would be diesels.

The influence of fuel prices on diesel sales is probably uncertain. Foreign countries with substantially higher real fuel costs do not exhibit high levels of dieselization (see Table 5.4). In the United States, real fuel prices are still far below those in most foreign countries. In 1980, the U.S. price of unleaded gasoline averaged $1.22 per gallon, while in Britain the price was $2.61 for the equivalent of a U.S. gallon and in France and Italy $3.13.

MANUFACTURING FACILITIES AND REGULATORY UNCERTAINTIES

For diesels to make up as much as 20 to 25 percent of new car and light truck sales in 1990, investment will be required in new or revitalized manufacturing facilities. Although some of today's diesel engines are modified versions of existing gasoline engines, most diesels are likely to be based on new designs that make optimum use of the special characteristics of diesel engines and the vehicles for which they are planned.

Major components must be manufactured specifically for light diesels, such as the engine block and head, camshafts, crankshafts, pistons and piston rods, injector pumps, valve lifters, and drive components. An efficient production module will make these components at a rate of 70 units an hour. On a three-shift basis, the module would produce approximately 384,000 units per year. The initial investment cost of such a module is estimated at $400 million to $500 million (in 1980 dollars).

For diesels to constitute 18 percent of new light vehicle sales in 1990, approximately 3 million cars and small trucks would have to be produced. If the domestic manufacturers account for 80 to 85 percent

TABLE 5.3   Possible Future Diesel Vehicle Sales and Use.

	Number of Light-Duty Vehicles (Millions)		Diesel as Percentage of Light-Duty Vehicles					
			If 10% of Sales in 1995		If 25% of Sales in 1995		If 50% of Sales in 1995	
	Sales	In Use	Sales	In Use	Sales	In Use	Sales	In Use
1980	14.4	139.2	3.2	0.4	4.1	0.4	5.7	1.0
1985	15.4	149.0	5.4	1.4	11.1	2.6	20.4	4.8
1990	16.5	158.7	7.7	3.5	18.0	7.6	35.2	14.9
1995	17.5	168.5	10.0	6.0	25.0	14.1	50.0	28.0
2000	18.5	178.2	10.0	8.2	25.0	19.6	50.0	39.1

of the sales, they will be producing about 2.5 million diesel passenger cars and light trucks in 1990. Accordingly, approximately seven production modules would be needed by 1990, each operating at the maximum. Because it is unlikely that each of the modules will be producing at their optimum rate, eight production modules will probably be needed. This expansion will require a capital investment of $3 billion to $4 billion. This is a large sum, but it should be viewed with long-range perspective. First, the investment will be made over a decade--and some of it has already been made. Second, it will be a small portion of the total investment of the U.S automobile producers for new plants, equipment, and tooling.

Overall, although the capital outlay for diesel manufacturing facilities is substantial, it is unlikely that serious financial problems will arise solely from the need to pay for diesel production. Moreover, in the first few years U.S. car makers will depend on foreign companies for light diesel engines--Volkswagen, Isuzu, Peugeot, BMW, Mitsubishi, and Toyo Kogyo.

Economic uncertainty is a permanent fact of life for the motor vehicle industry. Since the start of this century some 2,000 automobile assemblers and suppliers in the United States have merged or gone out of business. Car models must be designed far in advance of actual sales, requiring substantial commitments for equipment and materials long before the manufacturer is able to predict the market. Changes in consumer tastes, economic conditions, and, as the automobile industry has painfully learned in the last few years, gasoline prices and availability can lead to massive shifts in demand for automobiles generally and for individual models in particular. In such circumstances some technolgical, engineering, or manufacturing actions may not succeed.

TABLE 5.4 Trends in Diesel Sales and Fuel Prices in the U.S. and Selected Foreign Countries

	United Kingdom	France	West Germany	Italy	Japan	United States
	Diesels as Percent of New Car Sales					
1972	0.09	2.4	3.4	0.8		
1973	0.07	2.0	3.7	1.2		
1974	0.08	3.3	4.6	1.6		
1975	0.05	4.5	4.3	2.5	n.a.	
1976	0.5	4.3	3.8	2.5	0.4	
1977	0.5	6.4	4.7	2.5	1.1	0.1
1978	0.2	6.5	5.9	4.5	1.5	0.4
1979	0.3	7.3	7.2	4.1	2.0	2.2
1980	0.4	8.9	7.4	6.8	n.a.	4.6
	Gasoline (and Diesel) Prices in 1980 U.S. Dollars					
1979	2.44	2.84	2.31	3.18	3.14	1.20
1980	2.61 (2.73)	3.13 (2.21)	2.54 (2.44)	3.13 (1.46)	3.14 (3.14)	1.22 (1.13)

Source: Motor Vehicle Manufacturers Association of the United States.

Uncertainties involving the environmental and health consequences of diesel particulates are a different order of concern to the automobile industry. The committee has made preliminary estimates of these consequences, but, as discussed in Chapters 3 and 4, a great deal is not known. More research is being conducted, and more information will be forthcoming in the next few years. At one extreme, diesel particulates might be found not to constitute serious environmental and public health hazards, so that regulation beyond current levels would not be necessary. At the other extreme, diesel particulates might be found to be potent carcinogens, so that the government might decide to limit or ban future diesel production, or to set such stringent controls on emissions that diesel cars are effectively banned.

This uncertainty differs markedly from the ones automobile manufacturers normally face. If an individual model does not initially appear to be popular with consumers, the company can try a different pricing or advertising strategy. If diesel production were to be strictly limited or banned, efforts to influence demand would not be available. Production by some or all modules would have to cease. If the investment that had been made in diesel production facilities could be quickly and costlessly converted to others uses, the industry would have little cause for concern. Although the diesel components might be adapted to gasoline engines (just as some current gasoline engines have been converted to diesels), the converted engines are likely to be too big and inefficient for the vehicles. They might be adapted to larger vehicles, and greater flexibility might be designed into the entire range of vehicles manufactured by a company to take this eventuality into account. Such flexibility is expensive. The adaptation would be costly in any event, and if adaptation were not possible, much of the special equipment for diesel manufacture would have to be scrapped.

Thus, weak markets or regulatory impediments for diesel cars and small trucks would not mean a total loss of the investment in diesel facilities--though perhaps as much as 50 percent of the capital outlay could be lost. Clearly, the uncertainties concerning future regulation are an important consideration for the industry.

REFERENCES

Carlson, W. L. (1975). Crash Prevention Model. Accident Analysis and Prevention. 11:137-152.

Keeney, Spurgeon (1977). Nuclear Power Issues and Choices. Report of the Nuclear Energy Policy Study Group. Cambridge, Mass.: Ballinger Publishing Co.

McDonald, R., and G. K. Ingram (1981). Diesel Car Regulations and Traffic Casualties, Report to the Diesel Impacts Study Committee, National Research Council. Washington D.C.: National Academy Press (forthcoming).

McNutt, Barry, H. T. McAdams, and Robert Dulla (1979). Comparison of EPA and In-use Fuel Economy Results for 1974-1978 Automobiles--An Analysis of Trends. SAE Technical Paper 790932.

Mela, D. (1974). How Safe Can We Be in Small Cars? Proceedings of the Third International Congress on Automotive Safety, Vol. II. San Francisco, CA.

Stewart, J. R., and J. C. Stutts (1978). A Categorical Analysis of the Relationship Between Vehicle Weight and Driver Injury in Automobile Accidents. Highway Safety Research Center, University of North Carolina, Chapel Hill.

6 PUBLIC POLICY ISSUES AND ANALYTIC METHODS

Under certain ideal conditions an unregulated market efficiently satisfies the wants of consumers. But when the production or consumption of a good creates certain side-effects or "externalities," government intervention may be warranted to correct the situation. Such circumstances are frequently termed market imperfections or "market failures" (Bator, 1958). Air pollution from automobiles is widely regarded as an example of such an externality. The consumer, in operating a motor vehicle for his own private needs, produces emissions that impose costs on the whole society. The social costs of automobile use are not taken into account in the consumer's private cost-benefit calculations.

The divergence between private and social costs is, in theory, a justification for corrective government action. But such a divergence does not by itself dictate the appropriate means of intervention. The committee recognizes that in many cases government interaction in the marketplace may constitute a "cure" more harmful than the "disease" itself.

There are three major externalities associated with motor vehicle operations. First, as the earlier chapters have explained, the tailpipe emissions of motor vehicles contain pollutants that affect both human health and environmental quality. Thus, automotive air pollution is a recognized "negative externality." On the other hand, the private benefits to individuals for their fuel conservation efforts (e.g., the decision to buy a car with high rather than low mileage per gallon) are less than the benefits to the society as a whole. As noted below, the price of oil in the market does not reflect its true cost to our society--the social gains from saving a barrel of oil being greater than the private gains from saving a barrel. Thus, automotive fuel economy constitutes a "positive externality."

Those who concentrate only on negative externalities claim that emissions from diesel cars should be curtailed because of their possible adverse effects on public health and environmental quality. Those concerned only with positive externalities argue that diesel cars should be encouraged because they help conserve oil, which improves the nation's common weal. A balanced social view requires that both externalities be considered and all important consequences weighed before deciding on the optimal policies for diesels.

92

A third externality may be at work in the area of safety. As a result of health, disability, and life insurance coverage, society may pay some or all of the medical expenses, rehabilitation costs, and survivors' benefits resulting from accidents. In addition, the society may operate less productively or lose great talent temporarily or permanently by premature deaths or accidents. Therefore, society has a stake in preventing accidents. When a consumer chooses to purchase a safer car, he may thus impose a positive externality on society.

A system of competitive markets yields a specific distribution of income, which may not be considered by the community to be socially desirable. Regulatory intervention in the operation of markets is one means our political system has chosen to influence the distribution of income. Such considerations also influence the choice among methods of intervention when it occurs. The main reason why it has been national policy to keep the price of oil to individuals below its social value--why, for example, large excise taxes have not been levied on imported petroleum or on gasoline--is the perception by successive political administrations and Congresses of the consequences of such actions for income distribution. Thus, because this consideration forecloses higher fuel taxes, it is a matter of social policy to turn to such measures as fuel-economy standards and fuel-efficient diesels as ways of conserving oil.

CRITERIA FOR MAKING REGULATORY DECISIONS

At best the presence of externalities only indicates the direction in which public policy should move. It does not suggest the types of measures that should be taken or the magnitude of such measures. When both positive and negative externalities are present, a presumption as to direction may not even be possible. Criteria are necessary for making informed public policy decisions about the direction and magnitude of programs.

The overriding criteria in such decisions must be that society's resources are scarce--land, labor, capital, energy, and materials not being available in unlimited amounts. At the same time, multiple goals exist within our society--e.g., achieving ever rising real standards of living, providing socially equitable distribution of income, attaining better environmental amenities, which may not be incorporated in standard of living measurements, and improving upon health and safety. Reducing or eliminating the nation's vulnerability to disruptions of petroleum supplies is another worthwhile goal. Because the goals are multiple and open-ended and the resources are limited and costly, it is important to be efficient in the use of resources devoted to the various goals. Because the goals represent multiple dimensions of national welfare that cannot readily be measured, it is important to try to determine the appropriate tradeoffs among them.

COST-BENEFIT AND COST-EFFECTIVENESS ANALYSES

Cost-benefit analysis is an important analytical tool for comparing alternative uses of our nation's resources. The costs and consequences of public policy programs or proposals can be analyzed and compared; choices to achieve the greatest net social benefits can be made.

In cost-benefit analysis the social costs of a project or program are compiled and arrayed, and the social benefits are similarly compiled and arrayed. The benefits are measured, or inferred, using a variety of techniques based on the amount the consumer or beneficiary would be willing to pay for the goods or services provided by a specific activity. The difficulties in estimating the values are often considerable, especially if the goods or services are health and safety, because there is no direct way to purchase these in the marketplace. But the principle is the same: Willingness to pay is the most reliable value of a benefit. If the costs and benefits extend over more than one time period, a social discount rate should be used to compare values in different time periods. The same procedures should be done for all reasonable alternatives (including such things as a slightly more extensive or less extensive project). The alternative with the largest net margin between social costs and social benefits can perhaps then be identified.

Cost-benefit analysis can involve considerable methodological problems. The complexity results from efforts to determine the values of the social benefits, the social costs, and the discount rate, and in dealing with uncertainties in such values. Questions of who receives the benefits and who bears the costs are often involved. Greater complexity is introduced when measures that are not usually converted into monetary terms (e.g., air quality deterioration factors, mortality rates, and morbidity effects) are introduced into an evaluation.

Cost-effectiveness analysis is a limited form of cost-benefit analysis. It is designed to avoid the difficulties of comparing different types of benefits when they are not in common dollar units. For example, when the benefits of a particular government policy involve both savings of lives and gains in fuel efficiency, a complete cost-benefit analysis would require converting these two types of benefits into commensurable units. In cost-effectiveness analysis, on the other hand, the decision-maker compares only the costs of alternative policies, holding benefits fixed. The objective is to determine which of several policies achieves the same savings of lives at minimum cost.

Cost-effectiveness analysis is useful for discarding inefficient programs and for arraying the remaining efficient programs in a succinct manner. However, it cannot guide the choice among efficient programs, and it cannot guide the choice of targets. The efficient array can only provide a basis for decision-makers to make informed choices. The analysis leaves to decision-makers or the political process the task of determining the appropriate tradeoffs among various social and political goals--e.g., placing the perceived values of society on environmental amenities and on health and safety.

METHODOLOGICAL ISSUES IN COST-BENEFIT ANALYSIS

In performing a cost-benefit analysis of diesel particulate
regulations, the committee has had to consider a number of issues
involving the measurement and valuing of premature deaths as well as
injuries and illnesses attributable to pollution, the degradation of
air visibility, the cleanliness of the environment, and the
conservation of oil.

The Value of Preventing Early Deaths

The question of determining the value of extending a life or
avoiding a premature death is controversial, but it may play an
important role in the diesel decision. Life is not free of risk.
People are constantly exposed to risks of disease, injury, and death
from a variety of causes: industrial falls, say, or automobile
accidents and workplace contaminants, and such personal habits as
cigarette smoking and excessive drinking. Each of these risks is
usually of low probability for any individual. Among 220 million
people in our society, it stands to reason that even events of low
probability sometimes happen. There are 50,000 deaths annually from
traffic accidents and 100,000 deaths each year from lung cancer.

Public policy programs often involve purposeful changes in the
probabilities--e.g., reducing the probability of traffic fatalities or
decreasing the probability of lung cancer deaths. When extrapolated
across society, the programs mean quantifiable changes in the number of
early deaths. In advance of instituting a program, the deaths are
"statistical" and unidentified--the probability of a premature death
per person times the number of individuals affected.

The implied value that individuals put on small changes in the
probability of their premature death can be calculated from their
behavior with respect to risk. A number of empirical efforts have been
made in the past decade to measure the implied value of a statistically
premature death--the value of, say, a 0.1 percent change in the
probability of an early death times 1,000 people who might be
affected. These efforts have recently been summarized by Bailey
(1980). After making a number of adjustments to the reported findings
(e.g., he tries to separate the wage premiums that might be due to
increased probabilities of injuries from those due to increased
probabilities of early deaths), he attempts to place a value on
prolonging lives in a population at risk. This is termed the "value of
a life saved," a formulation representing the amount of benefits
obtained by an individual whose risk of death is reduced. Bailey finds
that the value of a statistical premature death probably falls in the
range of $170,000 to $715,000 in 1978 dollars. In 1980 dollars, this
would be a range roughly of $200,000 to $850,000. Freeman (1979),
examining the same data, decided to use a figure of $1,000,000 for the
purposes of evaluating the health benefits of EPA's pollution control
program.

Questions can be raised about the validity of these estimates. The estimates assume that the individuals involved are knowledgeable about the risks, that the sample is representative of the risk evaluation of the general population, that other factors in wage determination have been properly controlled, and that the wage premiums due to probabilities of injuries have been properly subtracted. Moreover, the willingness to pay for improving the chance of survival depends on the individual's income and attitude toward risk taking, as well as baseline or everyday susceptibility to risk.

All things considered, the use of a measure like the "value of a life saved" can be, despite the limitations, a guide to decision-making. For example, a goverment program that holds out promise of reducing premature deaths at a cost of $20,000 per individual is clearly desirable and relatively inexpensive. By contrast, a government activity that pays as much as $10 million to prolong the life of an individual may be considered a bad bargain, particularly if there are plenty of alternative goverment programs that could prevent early deaths at lower costs.

The Value of Avoiding Injury and Illness

In principle, the same data used for valuing life could be used to estimate the value of a change in the probability that individuals will incur a specific nonfatal injury or illness. The medical costs of treatment (and lost wages during treatment) are usually used as estimates of the value of avoiding injuries or illnesses. Use of the cost approach is, in the end, an underestimate of the social value, because, presumably, individuals would be willing to pay something more than the costs of medical care to avoid injuries or illnesses. The underestimate is likely to be more serious for permanently disabling injuries or illnesses than for completely reversible ones.

Sometimes the behavior of individuals--for instance, in selecting and operating cars--generates risks of injuries or death for themselves and for others. Individuals make different kinds of choices about exposing others to risks than they do about their own exposure to risks. In another National Research Council report, Chauncey Starr (1972) puts it aptly: "As one would expect, we are loath to let others do unto us what we happily do to ourselves." This assumed difference is, after all, the basis for most of our health and safety legislation.

Decisions on diesels might produce both types of injuries. As was discussed in Chapter 5, the wider use of diesels could result in fewer deaths and injuries in automobile accidents as a result of the use of heavier cars. Most of the reduced injuries from diesel use, or increased ones from limiting diesels, involve private consequences-- i.e., they are borne by the vehicle user.

By contrast, almost all health damage caused by automobile emissions is inflicted on others. Looked at more closely, some of the difference between these two categories blur. As a practical matter, society regulates some forms of behavior where the benefits are arguably predominantly private--e.g., prescription drugs. Moreover, given our predominant system of third-party payment for health care, others in society pay when individuals inflict injury on themselves (and families of the victims are affected in many other ways).

The Value of Visibility

Visibility, an important environmental amenity, may be reduced by the effect of widespread operation of diesel vehicles on ambient air quality. Some people will travel great distances to sightsee, backpack, or carry out other activities in clean air; some will pay high rent for apartments on top floors with vistas of the city, the sea, or the landscape.

These examples suggest possible ways of inferring how much people are willing to pay for visibility--i.e., what dollar value they put on it. Unfortunately, no direct inferences from actual behavior, comparable to the wage-risk empirical work mentioned above, have been drawn. A number of survey experiments have been conducted, in which individuals are shown photographs of comparative degrees of visibility and asked how much they would be willing to pay to avoid or achieve specified changes in visibility (Rowe, 1980). The committee draws on some of this work in Chapter 7.

The Value of Environmental Soiling

Higher levels of ambient particulate matter that would accompany more diesel-powered vehicles on the roadways would almost certainly increase the soiling of the environment. Differences between the properties of diesel particulates and those from other sources complicate the assessment of the potential adverse effects on vegetation, as well as buildings and other property. Indeed, there are grounds for believing that the oiliness of diesel particulates suspended in the atmosphere might cause a more serious soiling problem than an equal weight of particulates from other sources. The incremental contribution of diesel exhaust to soiling the interior of buildings, on the other hand, must be measured against a background level that may be considerably more intense with regard to soiling.

The committee has encountered a number of problems in attempting to make quantitative estimates of the level of soiling from diesel particulates and the costs that might result. No relatively simple and reliable measure of soiling could be found to serve as an index to the social costs involved. Nor could any suitable analysis be made for how an increase in the atmospheric loading of diesel exhaust might affect soiling.

The costs associated with soiling, including both the aesthetics of a grimier environment and the burdens of cleaning property, are difficult to calculate. While soot is one factor contributing to the need for repainting, for instance, it may also reduce the effects of weathering, another factor. The net effects have not been measured. Existing analyses of the costs of soiling do not lend themselves to

calculating the costs associated with various levels of diesel
particulates for this and other reasons. Valuing the consequences of
diesel-caused soiling, as it happens, was one of several shortcomings
in this analysis of diesel particulate regulation and will need to be
dealt with in any subsequent examinations of the issue.

The Value of Conserving Oil

Much more public policy in the past eight years has been devoted to
efforts to conserve petroleum than any other domestic natural
resource. From 1971 to 1981 domestic prices for domestic crude oil at
the wellhead were controlled, as were the margins of refiners,
distributors, and retailers of gasoline and diesel fuel. An important
consequence of such controls on fuel supplies is that prices for
gasoline and diesel fuel were probably below their true social
opportunity cost--the value to society of conserving oil. Consumers
did not have the incentive to pursue opportunities to conserve oil.

Public policy on petroleum supply has changed, and the wellhead
price of domestic crude is now linked to world market levels. Even
with domestic prices reflecting world market levels, there is still a
strong argument that the social opportunity costs of imported petroleum
is higher still. Dependence on foreign suppliers for 30 to 40 percent
of the oil used in the United States carries with it risks of
disruptions and consequent constraints on the independence of the
nation's political actions. The oil exporting countries are able to
exploit their market power and raise their prices from time to time
with serious economic consequences for the United States and other
nations. Limiting petroleum imports by means of, say, a tariff on
foreign oil, could reduce some of these costs. Lowering the U.S.
demand for imported oil may cause the price to drop, resulting in
additional savings on foreign petroleum. Acting jointly with other oil
importers would cause this effect to be even greater. Reducing oil
imports would improve the balance of payments, cause the dollar
exchange rate to appreciate, and allow a terms of trade gain vis-a-vis
our trading partners.

For any or all of these reasons, then, the value to society of
conserving an additional barrel of oil is greater than the market price
paid by consumers, even if the market price is at the world market
level. An obvious way to estimate this disparity is to calculate the
import duties on petroleum that would raise its price to the
appropriate social opportunity cost. Such an estimate is extremely
complex and involves many uncertainties. It is therefore impossible to
estimate this value with precision, and attempts to estimate the
difference between the social opportunity of oil and its market price
inevitably involve a wide range of uncertainty. Hogan (1981) has
estimated such an "import premium" at $2 to $40 for each barrel of
oil. His estimate implies that petroleum products are underpriced in
the market by $0.05 to $1.00 per gallon.

A consequence of the disparity between the consumer price and the
nation's social cost of oil consumed is that motorists undervalue the

social gains from making choices in favor of fuel economy. Even though
diesels reduce the cost of travel per mile and therefore are likely to
lead motorists to drive more miles. Such increases would not offset
the gains from diesel fuel economy. The monetary gains shown in Table
5.1 are probably, on balance, an underestimate of the value of the
social gains from greater fuel economy through the use of diesels.

Apart from the possible increase in miles driven, some of the
potential fuel economy gains from diesels are likely to be taken in the
form of purchases of larger vehicles than consumers would otherwise
buy. Although the national goal of pursuing petroleum conservation is
important, public policy recognizes it as one of many limited social
goals and not one that is intended to replace this country's commitment
to free consumer choice completely.

UNCERTAINTY

Virtually all of the entries in a cost-benefit calculation are
forecasts or estimates of statistical magnitudes, and all involve some
possible errors. The errors can be in the quantities: How many diesel
cars will actually be bought in 1990? What volume of particulates will
a 1990 diesel emit? Or the errors can be in prices or values: What
will be the price of gasoline in 1990? What is the pertinent value per
life saved? The arithmetic of cost-benefit analysis involves
uncertainties in the estimates and differences in the values of those
performing the analysis.

Risks due to uncertainties cannot be avoided in most decisions, but
they can be clearly specified and understood. Moreover, many of the
important decisions involving the use of diesel engines will be made
during the course of the next decade and beyond, not only in 1981 or
1982. During this period it is more than likely that new and better
information will be forthcoming on health and environmental effects and
consumer preferences in connection with diesel cars. This suggests the
utility of decision analyses that take explicit account of the
uncertainties and postulate future dates at which choices might be
made. Such an approach is an appropriate public strategy for the
review and regulation of diesels that takes account of the state of
information likely to exist and the range of consequences likely to
result when decisions are made. This type of analysis should be done
periodically.

This last method is the most sensible way of dealing with
uncertainty and is employed in Chapter 7.

REFERENCES

Bailey, Martin J. (1980). Reducing Risks to Life: Measurement of the
    Benefits. Washington, D.C.: American Enterprise Institute.
Bator, F. M. (1958). The Anatomy of Market Failure. Quarterly Journal
    of Economics. 72:351-379.

Freeman, A. M. III (1979). The Benefits of Air and Water Pollution
    Control: A Review and Synthesis of Recent Estimates. Council on
    Environmental Quality. NTIS Number A-NPB80-178759.
Hogan, William W. (1981). Decision Analysis of Regulating Diesel Cars.
    Report to the Diesel Impacts Study Committee, National Research
    Council. Washington, D.C.: National Academy Press (forthcoming).
Rowe, Robert D., Ralph d'Arge, and David S. Brookshire (1980). An
    Experiment on the Economic Values of Visibility. Journal of
    Environmental Economics and Management. 7(1):1-19.
Starr, Chauncey (1972). Benefit-Cost Studies in Sociotechnical
    Systems. In Perspectives on Benefit-Risk Decision Making. Report
    by the Committee on Public Engineering Policy. Washington, D.C:
    National Academy of Engineering, p. 30.

7 ANALYSIS OF CONTROLLING DIESEL CAR PARTICULATES

In this chapter the committee attempts the difficult task of assessing the aggregate risks and benefits of dieselization in the United States as a guide to public policy. One reason for the difficulty is that scientific evidence concerning the nature and scope of health and environmental risks is limited or inconclusive. The analysis is complicated by other factors: uncertainties about technological changes in operating efficiencies and emission controls for gasoline and diesel engines; preferences of the motoring public; prices and availability of petroleum-based fuels; various marketing and design strategies of automobile manufacturers; and the future course of public policy relating to risk-reducing regulations covering health, safety, and the environment.

Moreover, problems arise in valuing and comparing different types of private and social costs and benefits. For instance, social goals, such as maintaining a clean environment, and economic goals, such as lowering the inflation rate by reducing the quantity of oil from abroad, are difficult to compare. As another illustration of the problem, comparisons of similar appearing effects like cancer risks from emissions and death risks from traffic accidents involve different populations at risk over different time periods. Increased cancer risks affect older people predominantly and traffic risks affect mainly younger people. There is no simple way of resolving this difficulty. The committee's Analytic Panel has concluded that increased longevity is a better measure of the benefits of reductions in death rates than fatality statistics. In the analysis that follows, the concept of "person-years of life saved" is used to compare health and safety risks.

Although some analysts believe that the integration of dissimilar effects can be usefully achieved in principle, there are many alternative conceptual bases for such integration with different consequences. Almost always serious empirical problems occur in making such comparisons. Nonetheless, society somehow makes choices about such matters. Indeed, policy decisions concerning diesel cars are likely to be determined by public perceptions of risks and benefits at least as much as by conclusions drawn after rigorous analysis of scientific, technical, and economic data. For its part, the committee has attempted to facilitate the social decisions by displaying the various costs and benefits, and by offering a reasonable basis for their comparisons.

STRUCTURE OF THE ANALYSIS

The following benefit-risk analysis is structured to illustrate one method of evaluating and comparing the benefits and risks of regulating particulate emissions from diesel cars and small trucks. The uncertainties and complexities identified in previous chapters suggest the need to narrow and simplify the "regulatory problem" in order to make the analysis feasible. In its simplification, the committee assumes that government regulators have a choice between only two alternatives:

1. Accepting EPA's current limit on particulate emissions of 0.6 g/mi for diesel cars and light trucks in the model years 1982-1984 and 0.2 g/mi thereafter, or
2. Relaxing the standard for model years after 1984 by retaining the 0.6 g/mi maximum for particulate emissions from diesels.

To be sure, decision-makers face a wider range of choices. Even so, the detailed analysis of this restricted set of options lays the groundwork for comparing other alternatives.

This chapter therefore addresses the question: What is the balance of costs and benefits to society by shifting from a 0.6 g/mi to 0.2 g/mi particulate standard for diesel cars in model year 1985? If there are net benefits, then a standard below 0.6 g/mi, possibly 0.2, is preferable; if the costs outweigh the benefits, a standard at or above 0.2 g/mi may be preferable.

For purposes of this analysis, the 0.6 g/mi maximum is viewed as a baseline standard that is unlikely to be relaxed no matter what new information, additional analysis, or legislative change come forward. Manufacturers currently plan a number of modifications to vehicles and power systems for reasons of fuel efficiency and non-particulate emissions control, and the added cost of limiting diesel particulate emissions to 0.6 g/mi is estimated to be quite small. Coupled with the possibility that particulate emissions may threaten human health and damage environmental quality, the 0.6 g/mi standard appears to be a plausible and prudent baseline. Before formally concluding that this argument justifies an emissions standard of 0.6 g/mi, or less, it would be necessary to examine alternative ways to achieve the given level of emissions at lower social and private costs. The committee has not done this, however.

The 0.2 g/mi standard is the lowest level that may be technologically and economically feasible for all light-duty diesel vehicles to achieve in the near future. The assessment of emissions control technology in Chapter 2 indicates that achieving a 0.2 g/mi emission level will require a durable, reliable, and marketable particulate trap-oxidizer that may not be commercially available in 1985. A more stringent standard than 0.2 g/mi would probably effectively prohibit some types and sizes of diesel cars and small trucks. Although a particulate control technology is not yet proven, the adoption of a stringent emissions standard and a firm regulatory posture may still be supported as essential to stimulate the auto

industry to develop and market new emissions control technologies. In
such a situation the proposed standard is meant to "force" both the
industry and the government to advance the technology. The final
regulatory standard, of course, must be technologically feasible and
implementable, unless the intent is to prohibit the sale and use of
some types of diesel vehicles as a matter of public policy.

Several more sophisticated regulatory alternatives--most notably a
strategy for modifying standards as uncertainties about diesel impacts
are reduced or using standards that allow some averaging over the
fleet--are discussed at the end of this chapter.

In view of the conceptual and empirical problems of aggregating and
comparing dissimilar costs and benefits, the disaggregated specific
effects of diesel vehicles are presented first to determine whether any
conclusions can be reached without resolving the more difficult issues
of an integrated comparison of effects. An integrated comparison is
also presented, although interpretations of the figures should be made
with caution.

TREATMENT OF UNCERTAINTY

In comparing the 0.6 and 0.2 g/mi strategies, the Analytic Panel
considered how the costs and benefits of the emission levels vary as a
function of uncertainties about the use and consequences of diesels.
It turns out that a continuum of uncertainties appear in the factors
that influence diesels. The panel has estimated the likely bounds of
the factors and used the bounding values to examine the impact of
uncertainty on the estimate of benefits and risks in a small number of
situations. The situations are depicted in Figure 7.1, which
illustrates how uncertainties can compound into a complex of future
possibilities. The evaluation of uncertainty uses three sets of
cases: two sets of cases that use subjective bounds on the uncertain
variables that are analyzed--one using assumptions about uncertainties
that tend to maximize the benefit of adopting the 0.2 g/mi standard,
the other using assumptions that maximize the benefit of the current
0.6 g/mi standard (that is, minimize the benefit of shifting to 0.2
g/mi). A third set of cases is also analyzed to show the effects of
less extreme assumptions made at an intermediate level between the
upper and lower bounds, though the confidence that should be placed in
the results of this set of cases is not clear. The intermediate set of
cases cannot be viewed as "most likely" because the uncertainties
associated with diesel impacts are so great that the identification of
a single "most likely" case is difficult and misleading.

The analysis of the two extreme sets of cases indicates how
sensitive the choice of standards is to the range of uncertainties and
shows whether one standard dominates the other for the entire range of
assumptions. If it can be demonstrated that the shift to the 0.2 g/mi
standard is not beneficial, even under assumptions maximizing the
benefits of that standard, then some standard greater than 0.2 g/mi is
optimal--possibly 0.6 g/mi. Similarly, if the 0.2 g/mi standard is
beneficial, even using assumptions that maximize the benefits of

Figure 7.1   Decision Flow of Considerations in Evaluating Diesel Emission
             Standards

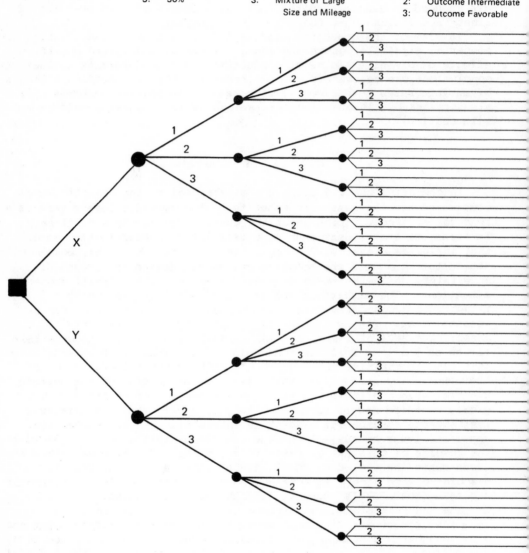

POST-1984
EMISSION STANDARD
DECISIONS

Option X: 0.6 g/mi
Option Y: 0.2 g/mi

LONG-TERM
MARKET SHARE
OF DIESEL CARS

1:   10%
2:   25%
3:   50%

CONSUMER PREFERENCES
FOR DIESEL CARS

1:   High Mileage
2:   Large Size, Luxury,
     and Safety
3:   Mixture of Large
     Size and Mileage

EFFECTS OF REGULATION
ON OIL IMPORTS, HEALTH,
ENVIRONMENT, AND
OPERATING COST

1:   Outcome Unfavorable
2:   Outcome Intermediate
3:   Outcome Favorable

■ "Decision" Node
● "Chance" Nodes

retaining the 0.6 g/mi standard, then a standard below 0.6 g/mi is preferable.

The analysis of the two extreme sets of cases can also be used to evaluate the maximum possible losses or "regrets" that are apt to result from selecting the wrong standard. For example, the maximum regrets from irrevocably selecting a 0.2 g/mi standard would be the losses incurred if the actual circumstances turned out to be those that maximized the benefits under the 0.6 g/mi standard. Similarly, the maximum regrets from the selection of an irrevocable 0.6 g/mi standard would be the losses when actual circumstances strongly favor 0.2 g/mi. The regrets calculations indicate the down-side risks that policy-makers face in selecting either standard. In addition, the regrets calculations suggest the magnitude of the maximum potential gains from a strategy of sequential decisions about standards, phased so that new information about uncertainties can be considered as it becomes available. The maximum regrets are the losses that are avoidable if the decision can be made on the basis of valid information.

The use of regret analysis, the panel recognizes, is only one of many possible methods of decision-making and one whose weaknesses have been discussed in the literature on decision theory. Regret analysis points to the policy that minimizes the worst outcome. The panel did not attempt to apply an alternative method. In particular, it did not assess the probabilities of each outcome, either subjectively or objectively, and then calculate the expected costs and benefits. Thus, the conclusions are especially sensitive to the extreme values chosen to represent the worst cases. Since the panel's assessment of the worst cases necessarily contains subjective elements, the specific numerical results should be viewed with caution. This methodology is sometimes called the "minimax" regret principle of decision-making. It seeks to examine the advantages and disadvantages as well as the alternatives.*

FORECASTING THE LIGHT-DUTY VEHICLE MARKET

Assumptions about the characteristics of the light-duty vehicle market and its reaction to various particulate emissions standards obviously also affect the evaluation of the 0.6 and 0.2 g/mi standards. Especially important are the share of the market that is equipped with diesel engines and the average weight and fuel economy of the fleet.

To simplify the analysis, the effects of 0.6 and 0.2 g/mi standards are estimated only for the year 1995. The fleet for the 1995 model year was selected because the diesel share of sales of new light-duty vehicles can reasonably be assumed to have leveled off by then, and the full impact of regulatory decisions made in the 1980's will be felt.

---

*For a discussion of minimax regret analysis, as well as an exposition of alternative decision methods that do not require the explicit assessment of probabilities, see Robert Duncan Luce and Howard Raiffa. Games and Decisions. John Wiley and Sons, Inc., New York, 1957.

The future market share of diesel cars depends on the relative cost and performance of diesel and gasoline engines and the relative values that car buyers place on such factors. Today's diesel cars have a slightly higher initial purchase price, higher maintenance costs, and lower average fuel costs than comparable gasoline vehicles. As explained in Chapter 5, such factors result in a 2 to 4 percent discounted cost savings for the average driver over the life of the car (provided the car's life is 100,000 miles). Even larger savings are achieved by motorists who use their cars more extensively or drive more in city traffic (where the diesel's relative fuel efficiency is high). Diesel vehicles exhibit such performance problems as difficulty in cold starting, slow acceleration, engine chatter, and exhaust odor that partly offset the appeal of their lower long-term cost. Even so, the diesel's small but increasing share of new vehicle sales implies that many motorists are attracted by the economies they offer. The committee's analysis suggests that diesel cars might eventually capture as much as 25 percent of the market for light-duty vehicles at current relative price and performance levels, although much lower or higher shares are also plausible.

The relative cost and performance of diesel engines may change in the future, giving rise to further questions about the diesel's future. On the one hand, rapid increases in real fuel prices or manufacturing or technological improvements that reduce the purchase price, lower the maintenance costs, and eliminate the performance problems of the diesel could increase its share of the automobile market. On the other hand, stable or declining fuel prices or rapid improvements in gasoline engine fuel economy could depress the diesel's market share.

Our baseline assumption is that the diesel's share of the light-duty vehicle market will be 25 percent of sales by 1995 at the 0.6 g/mi particulate standard--a sales level forecast by some car makers.

EFFECTS ON NEW VEHICLE SALES

The principal market impact of the costs of emission controls to meet the 0.2 g/mi standard in model year 1985 might be to change the pattern of vehicle sizes and vehicle efficiencies. The two characteristics are obviously closely related. To predict the changes in passenger cars requires an understanding of how consumers might react to various competing attributes of different vehicles when making their purchase decisions, as well as forecasts of other relevant factors, such as future fuel prices, the government's new car fuel economy standards, and the ability of auto manufacturers to develop fuel efficient "heavy" cars, as well as their relative preferences for large and fuel efficient vehicles. Obviously, many of the future developments in fuel prices and fuel efficient automotive technologies are difficult to predict.

Unfortunately, history provides little guidance to the relative value that consumers place on these two factors, largely because real gasoline prices have been fairly stable until recently. Thus there was little incentive to trade off weight for fuel economy. U.S. car buyers purchased relatively heavy automobiles through the 1950's and 1960's, but this choice may have reflected increases in real income and decreases in real gas prices, rather than changing preference for larger and heavier cars. Average new car fuel economy increased and weight decreased rapidly during the 1970's, especially after the surges in real gas prices during 1973-1974 and 1979-1980. The changes in the 1970's suggest a strong willingness by consumers to sacrifice weight for fuel efficiency, caused mainly by the sudden and unexpected gas price increases, coupled with the fear of gas lines or rationing, leading some buyers to overreact. If gasoline prices stabilize or increase at steady, predictable rates, reducing anxiety about fuel availability, some motorists may return to relatively heavy cars.

Recent research suggests that the strength of car buyers' preferences for heavier cars will probably depend on the ability of auto manufacturers to incorporate luxury features commonly associated with heavy cars into light cars at reasonable costs. Buyers of heavy cars do not appear to value weight per se, but rather other attributes that have been traditionally found in heavy cars, such as comfort, spaciousness, safety, greater reliability, smoother ride, reduced interior noise, air conditioning, and elegant or flashy features such as body styling and interior upholstery. Many of these features can probably be more cheaply provided in a heavy car than in a light car, though in some cases the cost differential may not prove large. The proliferation of small, light cars that provide luxury features previously available only on large, heavy cars during the 1970's suggests that manufacturers are designing light cars that appeal to many former buyers of heavy cars.

The most critical assumptions about the light-duty vehicle market are the changes that occur with the imposition of a 0.2 g/mi standard rather than a 0.6 g/mi standard. As indicated in Chapter 2, achieving the 0.2 g/mi standard will require trap oxidizers for all but some of the very small diesels. The fleet average cost is likely to be between $150 and $600 per vehicle. The resulting increase in diesel life-cycle costs may induce some potential diesel owners to switch to gasoline (or small diesel) vehicles, with subsequent impacts on fuel consumption, traffic safety, consumer costs, and public health.

The response of car buyers is an important determinant of the benefits of shifting to the 0.2 g/mi standard. If the market share of diesel cars does not fall when the cost of the emission control device for meeting the 0.2 g/mi standard is added to the "sticker price," there should be no changes in road mortality and fuel consumption. By contrast, in the extreme situation where the cost of a tighter standard reduces diesel car purchases to zero, consumers will suffer economic losses, traffic fatalities will increase, and fuel consumption will rise--the exact sequel depending upon the type of gasoline vehicle bought by would-be diesel owners. If motorists substitute gasoline-powered automobiles of identical weight but lower fuel economy for the diesel car they would have bought under the 0.6 g/mi standard, oil imports should increase but traffic fatalities should not be affected. By contrast, if motorists substitute gasoline vehicles of identical fuel efficiency but lighter weight than the diesel cars replaced traffic deaths should increase but oil imports should not.

The market reaction to additional particulate control costs cannot be forecast with much confidence. To date, experience with diesels is too limited to forecast the likely difference in its share with any precision. Similarly, the relative preferences of car buyers for larger, heavier cars versus fuel economy are not sufficiently understood to accurately predict whether potential diesel owners would choose gasoline vehicles with similar weight or, instead, with similar fuel economy.

Given the uncertainty about the effect that particulate controls will have on vehicle choices, three assumptions about market response are examined in this analysis (see Table 7.2). Assumption 1 holds that the diesel's market share is unaffected by tighter controls. Assumption 2 posits that the diesel's market share is affected and potential diesel buyers switch to gasoline cars of the same weight. Assumption 3 puts it that the diesel's share is affected and that potential diesel buyers change to gasoline cars with the same fuel efficiency. The actual response of the car market is likely to involve some shift in diesel sales and a combination of the responses in the second and third cases.

MEASUREMENT OF COSTS AND BENEFITS OF DIESEL EMISSIONS CONTROLS

The comparison of the 0.6 and 0.2 g/mi alternatives requires a calculation of the costs and benefits (including both individual and societal) of regulating emissions. Quantitative estimates of costs and benefits in the area of diesel emissions are extremely difficult both for the reasons mentioned above and, more generally, for the lack of widely accepted measurements or projections for many critical values. For these reasons the committee has not attempted to make definitive quantitative estimates. For purposes of clarity, the calculations are presented in detail, but, unless otherwise specified, they should only be interpreted as at best indicative of orders of magnitude.

Ambient Particulate Levels

An understanding of how health, visibility, materials damage, and other elements of environmental quality could improve by lowering diesel particulate emissions requires an understanding of the linkages between tailpipe emissions and ambient air quality. As discussed in Chapter 3, a description of this linkage requires many detailed assumptions and measurements that are geographically specific. Table 7.2 presents the EPA's estimates of ambient particulate levels based on data from the National Air Surveillance Network.

For purposes of estimating costs and benefits, the committee has developed projections of tailpipe emissions of diesel particulates from the future fleet of diesel cars. The level of aggregate emission reductions that would accompany a shift in the emissions standard in model year 1985 from 0.6 g/mi to 0.2 g/mi is presented in Figure 7.2, based on the assumption that diesel cars and light trucks will equal 25

TABLE 7.1 Regional Ambient Levels of Light-Duty Diesel
Particulates Corresponding to Nationwide Emissions of
150 to 250 Thousand Metric Tons*

Population Category	City	Particulate Level (micrograms per cubic meter) Light-Duty	Heavy-Duty
Over 1 million	Chicago	3.0 – 5.1	2.3 – 3.3
		6.3 – 10.7	4.9 – 7.0
	Detroit	2.1 – 3.5	1.6 – 2.3
	Houston	4.4 – 7.5	3.4 – 4.9
	Los Angeles	5.7 – 9.6	4.3 – 6.2
	New York	2.2 – 3.8	1.7 – 2.4
		2.8 – 4.8	2.2 – 3.1
	Philadelphia	2.6 – 4.4	2.0 – 2.9
	Average	3.6 – 6.2	2.8 – 4.0
500,000 to 1,000,000	Boston	1.9 – 3.3	1.5 – 2.1
	Dallas	6.4 – 10.8	4.9 – 7.0
	Denver	2.0 – 3.4	1.5 – 2.2
	Kansas City, MO	1.5 – 2.5	1.1 – 1.6
	New Orleans	2.2 – 3.8	1.7 – 2.5
	Phoenix	4.4 – 7.5	3.4 – 4.9
	Pittsburgh	1.8 – 3.0	1.4 – 2.0
	San Diego	2.4 – 4.0	1.8 – 2.6
	St. Louis	2.5 – 4.2	1.9 – 2.7
	Average	2.8 – 4.7	2.2 – 3.1
250,000 to 500,000	Atlanta	2.2 – 3.7	1.7 – 2.4
	Birmingham, AL	2.6 – 4.4	2.0 – 2.8
	Cincinnati	1.7 – 2.9	1.3 – 1.9
	Jersey City	2.2 – 3.7	1.7 – 2.4
	Louisville	2.0 – 3.4	1.5 – 2.2
	Oklahoma City	3.5 – 5.9	2.7 – 3.9
		2.1 – 3.6	1.6 – 2.4
	Portland	1.7 – 2.9	1.3 – 1.9
	Sacramento	2.2 – 3.8	1.7 – 2.4
	Tucson	1.6 – 2.7	1.2 – 1.7
	Yonkers, NY	2.4 – 4.1	1.9 – 2.7
	Average	2.2 – 3.7	1.7 – 2.3
100,000 to 250,000	Baton Rouge	2.0 – 3.3	1.5 – 2.2
	Jackson, MS	1.7 – 2.9	1.3 – 1.9
	Kansas City, KA	0.9 – 1.5	0.7 – 1.0
		1.3 – 2.1	1.0 – 1.4
	Mobile, AL	2.0 – 3.4	1.5 – 2.2
	New Haven	2.4 – 4.1	1.9 – 2.7
	Salt Lake City	2.1 – 3.5	1.6 – 2.3
	Spokane	1.2 – 2.1	0.9 – 1.3
	Torrance, CA	5.0 – 8.4	3.8 – 5.5
	Trenton, NJ	1.9 – 3.1	1.4 – 2.0
	Waterbury, CT	3.8 – 6.7	2.9 – 4.4
	Average	2.2 – 3.7	1.7 – 2.4
Under 100,000	Anchorage	2.1 – 2.7	1.6 – 1.7
	Helena, MN	0.6 – 0.8	0.5 – 0.5
	Jackson Co., MS	0.9 – 1.7	0.7 – 1.1
	Average	1.2 – 1.7	0.9 – 1.1

Source: U.S. Environmental Protection Agency, 1980c.

Figure 7.2 Projected Nationwide Emissions Reduction Resulting from Changing the Diesel Particulate Standard from 0.6 to 0.2 g/mi.

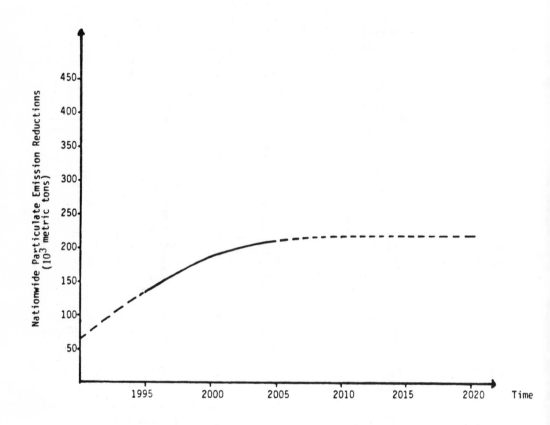

Diesel car market rises to a limit of 25 percent of total new light-duty vehicle sales in 1995.

percent of all new light-duty vehicle sales by 1995. Under such
assumptions, the 0.2 g/mi standard could reduce diesel emissions levels
by about 200,000 metric tons nationwide per year by the year 2000.
Using the EPA's ambient projections, the aggregate tailpipe emission
reduction could result in lowering ambient levels by about 5 $\mu g/m^3$
in highly polluted urban areas and by less than 1 $\mu g/m^3$ reductions
in cleaner urban and rural areas.

Effects on Health

In Chapter 4 and, with greater detail, in its report (National
Research Council, 1981a) the Health Effects Panel has stated that the
data collected so far on the adverse health consequences of diesel car
emissions do not provide a complete basis for risk assessment. Still,
the possibility that diesel cars may increase the incidence of lung
cancer is a cause for concern, though the potential for cancer in other
organs may also exist because inhaled particulate matter and bioactive
chemicals are systematically removed from the lungs to other organs.

The morbidity implications of diesels, which include all potential
non-cancer pulmonary and systemic health impairments described in the
report of the Health Effects Panel, could exceed the impact of lung
cancer that may be induced by particulates laden with hydrocarbons.
While there is a lack of usable quantitative information on which to
base an analysis of the implications of diesel-induced morbidity, the
importance of morbidity factors, such as the possibility that diesel
exhaust will exacerbate chronic obstructive pulmonary disease, should
be part of any future analyses of diesel risks and benefits when
adequate quantitative data become available.

In general, the diesel's contribution to total ambient quantities
of particulates in the atmosphere would be only a few percent.
However, the small size and chemical properties of the particles may
cause problems disproportionate to their total contribution to the mass
of suspended particulates in the air. Nevertheless, there is no sound
basis for estimating an incidence of pulmonary and systemic effects
from such low levels. In selected localities, however, such as the
narrow "street canyons" in business districts of major cities, diesels
may be highly concentrated, so that the level of diesel particulates
would be much greater. The possibility cannot be excluded that people
concentrated in such areas may suffer measurable pulmonary and systemic
effects from diesel exhaust.

Cancer Risks

The analysis performed for the committee by Harris (1981) of
available human epidemiological studies and nonhuman laboratory
research is useful to quantify the range of uncertainty of lung cancer
related to diesel particulates. (See also Chapter 4.) His analysis is
based on the following simple model:

$$\text{Relative risk of lung cancer} = 1 + r \times C \times D,$$

where C is the ambient concentration of particulates (measured in $\mu g/m^3$), D is the duration of exposure to diesel emissions (measured in years), and r is the parameter that must be estimated from available epidemiologic and laboratory evidence. Thus, the quantity C x D measures the cumulative exposure to diesel emissions (in $\mu g/m^3$ years). For example, an incremental exposure of 2 $\mu g/m^3$ over a 20 year period amounts to a value of C x D or 2 x 20 = 40 $\mu g/m^3$ years.

The relative risk of lung cancer represents the ratio of lung cancer incidence from a given diesel exposure to the lung cancer incidence without diesel exposure. Hence, in the models the absolute increment in lung cancer risks depends on the baseline risk of lung cancer from all other sources. That is, diesel particulates may multiply whatever risk of lung cancer is already present.

Based on the available epidemiologic and laboratory evidence, Harris estimates a 95 percent upper confidence limit of the parameter r to be 0.0005. This estimate of the upper confidence limit of increased risk of lung cancer does not provide an absolute measure of human health impact. The estimated lower confidence limit of r, it should be noted, includes the possibility of no effect on lung cancer risks.* The upper confidence limit does serve, however, as an indicator of the extent of uncertainty regarding the carcinogenic effects of diesel engine emissions in humans. Despite these limitations the upper confidence limit is considered significant in comparing the potential risks of the ambient population exposure to diesel engine emissions with other personal and societal risks.

From the formula above, a middle-aged man exposed to an average increment of 2 $\mu g/m^3$ of diesel particulates for 20 years would incur in the upper limit a relative risk of lung cancer equal to

$$1 + 0.0005 \times 2 \times 20 = 1.02.$$

or, equivalently, an increase in lung cancer risk in the upper limit by 2 percent. To place this in perspective, a male who had been smoking cigarettes for 20 years would incur a ten-fold to twenty-fold increase in lung cancer risk. A man exposed through his job to asbestos for a similar number of years incurs a two-fold to eight-fold increased risk of lung cancer.

For U.S. males currently aged 55 to 64, the annual lung cancer death rate is about 220 per 100,000 population, of which about 180 to 200 per 100,000 are attributable to cigarette smoking. Based on the above model, the effects on lung cancer death rates from exposure to an average increment of 2 $\mu g/m^3$ of diesel particulates for 20 years could range from no increase to an upper limit of 4 per 100,000.

---

*The statistical analysis performed by Harris indicates that the confidence interval of parameter r is -0.00025 to +0.0005.

Non-Cancer Risks

Diesel vehicles also pose health risks other than lung cancer. The gaseous and particulate emissions of diesel engines, like those of gasoline engines, may cause eye irritation, induce cardiac stress, increase pulmonary infection, and exacerbate other diseases. The key issue is whether a diesel vehicle is likely to cause more or less of these non-cancer health effects than the gasoline vehicles it replaces.

The quantity of particulates, aldehydes, and some other emissions from light-duty diesel engines are significantly higher than those from catalyst-equipped gasoline engines. It has been estimated that if diesels formed 25 percent of the light-duty fleet, aldehyde emissions would increase by about 15 percent (National Research Council, 1981b). Aldehyde emissions are of concern because they contain acrolein and formaldehyde, both known irritants. Formaldehyde is also a potential carcinogen. They also may promote oxidation of other air pollutants that could result in secondary aerosols.

The possible adverse health effects from higher particulate, aldehyde, and some other emissions may be offset, however, by lower hydrocarbon and CO emissions. Lower hydrocarbon emissions may reduce the formation of photochemical oxidants, which cause serious health problems for people with respiratory afflictions. Moreover, the low volatility of diesel fuel compared to gasoline results in a significant reduction in fugitive hydrocarbon emissions that escape from refinery storage facilities and vehicle fuel tanks and from the act of the refueling motor vehicles.

Hydrocarbon emissions from diesel vehicles are about 40 percent lower than from comparable gasoline vehicles. Consequently, if diesel cars and light trucks amounted to 25 percent of the whole fleet, they would reduce hydrocarbon emissions by at least 10 percent (Gray, 1980). The 10 percent reduction is considered a minimum because in-use deterioration of emission levels from gasoline vehicles is much greater than that of diesel vehicles.

Carbon monoxide emissions may cause health problems for persons with cardiovascular or peripheral vascular disease. Currently, all light-duty vehicles are required to meet a CO standard of 3.4 g/mi, although many waivers of up to 7.0 g/mi have been granted. Diesels emit CO at levels 33 to 50 percent below the 3.4 g/mi standard; consequently, if diesels powered 25 percent of the nation's cars and small trucks, CO would diminish 12 to 17 percent.

The possible non-cancer health effects of diesels are ignored in the committee's comparison of the 0.2 and 0.6 g/mi standards, largely because the lack of usable quantitative analysis of such effects prevents further analysis of the implications of diesel-induced morbidity. The omission of non-cancer effects may not seriously limit the analysis if gains from lower hydrocarbon and CO emissions offset the losses from greater particulate and aldehyde emissions. Still, it is possible that diesels, on net, will have health consequences other than cancer.

Effects on the Environment

Of the various environmental impacts described in Chapter 3, the one that has been analyzed most specifically is visibility. Reduced visibility, particularly in urban areas, is the most obvious effect of particulates. Table 3.2 illustrates the impact on visibility of potential diesel particulate concentrations in the Los Angeles area. To measure the importance of this visibility loss, the data from 15 metropolitan areas were extrapolated to all urban households and an attempt was made to estimate the value of this loss. Rural areas are presumed to be unaffected.

Attempts have been made to measure the value of reductions in visibility, but the valuation of the loss of an amenity that has no simple market-determined value is difficult. One experiment (Rowe, 1980), for instance, found that householders in the vicinity of Shiprock, a major natural monument in the Southwestern United States, were apparently willing to pay about $57 per year to avoid a deterioration in visibility of 75 to 50 miles. In another experiment in the Los Angeles area (Brookshire, 1979), householders were estimated to be willing to pay approximately $8 per month for the aesthetic component of a substantial improvement in air quality, meaning mainly visibility.

The data gathered by Rowe and his colleagues may be taken to imply that, on the average, the households in the sample valued changes in visibility at about $2.30 per mile per year. The Brookshire data cannot be used for comparison because it does not specify the quantitative change in visibility. As an upper limit, the committee has used a value of $6.00 per year for individual households for each mile of visibility change. Such values are probably not definitive but they suggest a range and order of magnitude for visibility loss. A reasonable lower limit for the value of visibility loss is assumed to be zero because in most urban areas a small worsening in average visibility caused by diesel exhaust might go unnoticed against a background of normal variations from other man-made and natural factors.

Suspended particulates in the air are a common source of soiling to the outer surfaces of buildings. In combination with corrosive pollutant gases and water, deposited particulates can create significant damage to structures and materials (National Research Council, 1979). While soiling damage is most easily attributable solely to the deposition of particulates, little empirical physical research or econometric measurement has been performed on this condition in the last ten years. (U.S. EPA, 1980a). This has made it impossible for the committee to develop a range of soiling effects or a measure of the harm caused.

Without more definitive understanding of how diesels contribute to soiling, then, the major environmental impact likely to be affected by decisions on diesel particulate standards is a reduction of urban visibility. Other environmental effects associated with diesel use are either not influenced by particulate regulations or cannot be assessed because not enough data are available. Atmospheric heating and secondary aerosols fall into this category.

Effects on Motorist Expense and Satisfaction

Because of their greater relative fuel efficiency, diesels offer lower ownership and operating costs than gasoline-powered cars of equal weight over the vehicle's life cycle. Table 5.2 shows the effect of diesels on lifetime costs for each weight category of cars powered by diesel and gasoline engines. In practice, some motorists might choose to replace gasoline-powered cars with diesels of equal lifetime costs--that is, they may buy heavier diesels. Or they may choose some combination of reduced lifetime costs and increased vehicle weight.

Particulate regulations may affect the benefits that motorists enjoy from diesels. Pollution control equipment required to meet a stringent particulate standard, such as 0.2 g/mi, will increase the cost of purchasing a diesel automobile or light truck and may contribute to additional operating and maintenance costs as well. The extra costs will induce some consumers who would have bought diesels at the 0.6 g/mi standard to buy gasoline cars when the 0.2 g/mi standard is in effect.

Each consumer who chooses to purchase a diesel car despite the added expense of the 0.2 g/mi standard suffers a loss equal to the price of the control equipment and additional operating costs, as well as the value of the additional inconvenience of maintaining the control equipment. Consumers who switch their purchases from diesel to gasoline automobiles to avoid the added expense also are worse off in terms of consumer satisfaction because they do not get the diesel they would have preferred. The basis for their preference might be any of a variety of reasons, including better fuel economy, driving range, engine durability, or some attitudinal factors such as the technological novelty of the diesel or the prestige of owning a car that is the subject of so much advertising and publicity. The value of the satisfaction lost in switching will be less than the additional lifetime cost (or they would have continued to choose diesels) but greater than zero (or they would have chosen gasoline cars originally). The average loss for those induced to switch by added costs of the 0.2 g/mi standard is assumed to be half the additional lifetime costs.

Effects on Traffic Safety

The effect of diesel cars and light trucks on traffic safety was discussed in Chapter 5. The proposed 0.2 g/mi standard could affect traffic safety if the cost of tighter controls induced some motorists to switch from diesels to small, lightweight vehicles that achieve comparable fuel economy but are less safe in collisions. Compact and subcompact cars tend to get into about the same number of accidents per vehicle mile as larger vehicles, when differences in average driver ages and other factors are controlled. But accident statistics compiled by the National Highway Traffic Safety Administration show that lighter cars contribute to many more fatalities and serious injuries per accident than large cars, again adjusting for other factors.

The future relationship between average vehicle weight and injury and fatality rates is uncertain, particularly because auto manufacturers may design more safety features into automobiles (in response to either consumer demands or government regulations) and small car drivers may behave more cautiously to compensate for the higher risks. Nevertheless, significant differentials in the fatality and injury rates are likely to persist. Future safety regulations are certain to make both large and small cars safer, but for now, the accident statistics indicate, drivers of small cars do not drive cautiously enough to compensate for the inherent vulnerability of their vehicles.

The committee's analysis considers the effects of the 0.2 and 0.6 g/mi standards only on traffic fatalities, not on serious injuries. This may be an important omission because injuries are much more numerous than fatalities and are also related to vehicle weight.

Effects of Reduced Fuel Use

If the 0.2 g/mi standard causes diesel cars to be displaced by gasoline cars of equal weight, the fuel efficiency of the entire fleet of passenger cars will decline and fuel consumption will increase. The increased demand for oil will be reflected, in turn, by an increased demand for oil imports. The extent of this increase depends critically on the on-the-road fuel efficiency differences beween diesel-powered and gasoline-powered vehicles. As discussed in Chapter 2, the fuel efficiency differential is uncertain and could change over time as a result of further technological developments. For purposes of this analysis, the committee has assumed that in the 1990's diesel vehicles will be 30 percent more efficient than comparable gasoline vehicles. It may be that the manufacture of diesel engines and the servicing of the vehicles is more energy intensive than gasoline vehicles, thereby effectively reducing the fuel advantage. The committee has not examined this possibility.

It is widely held that the cost to society of imported oil exceeds the market price to oil consumers for a variety of reasons having to do with economics, foreign policy, and national security. This belief is the basis for the CAFE standards and the "gas guzzler" tax. Attempts to measure the difference between private and social costs of oil have resulted in a wide range of estimates. The analysis of the problem by Hogan (1981) discussed in Chapter 6 presents a range of $2 to $42 per barrel of imported oil as the current value of the "import premium." To assess this effect for the mid-1990's a narrower range of values from $0.10 to $0.50 per gallon has been assumed in the committee's analysis.

BENEFIT-COST ANALYSIS OF A MORE STRINGENT STANDARD

The major uncertainties in assessing the relative merits of the 0.2 and 0.6 g/mi standards are those relating to:

- The adverse or beneficial impacts of any given level of dieselization--e.g., the carcinogenic potency of diesel particulates, the value of reduced visibility, the cost of trap-oxidizers, the degree to which traffic fatalities increase when passenger cars are downsized, and the social costs of oil imports; and
- The response of consumers to diesel cost increases associated with a shift to the 0.2 g/mi standard.

The level of diesel sales under a 0.6 g/mi standard is highly uncertain for reasons discussed above. However, while it does affect standards being compared, it does not affect their relationship to each other under the various uncertainties examined.

To avoid dealing with the uncertainties that affect the level of benefits equally under both regulatory standards, the results that follow will deal only with the differences in net benefits resulting from the EPA plan to adopt the 0.2 g/mi standard for post-1984 vehicles, in comparison to the alternative of continuing the 0.6 g/mi standard for those vehicles. To calculate the differences in net benefits between the two standards, the committee has analyzed nine different cases. The nine cases reflect three different market responses to the shift in particulate standards, cross-classified by three different assumptions about health, safety, and environmental impacts of dieselization. (See Figure 7.1 for a decision flow or decision "tree" diagram for this analysis.) As shown in Tables 7.2 and 7.3, Set A includes assumptions that maximize the net benefits of 0.2 g/mi and Set C includes assumptions that maximize the net benefits of retaining 0.6 g/mi, while Set B has intermediate assumptions. The panel derived the specific numerical values in Table 7.3 from the estimates of economic effects in Chapter 5 and supporting documents. Although the panel relied wherever possible on available objective information, it needs to be recognized that Table 7.3 contains subjective elements.

A summary of the results of the case where diesel sales under the 0.6 g/mi standard reach 25 percent is presented in Table 7.4. The committee calculated results for diesel sales levels of 50 percent and 10 percent, but the major conclusions of the analysis as illustrated by the 25 percent case remain unchanged at the higher and lower levels of diesel sales. The analysis of the differential benefits and costs of changing the emissions standard from the 0.6 g/mi to 0.2 g/mi is presented in five dimensions:

- Person-Years Saved by the decreased incidence of lung cancer as fewer particulates are emitted if the standard is strengthened from 0.6 g/mi to 0.2 g/mi;
- Value of Visibility Gained in urban areas as a result of decreased particulate emissions;
- Person-Years Lost from increased traffic fatalities caused by a reduction in average vehicle size as some consumers shift from diesel vehicles to smaller gasoline vehicles when diesel particulate control system costs increase;

TABLE 7.2 Cases Used in Analyzing Net Benefits of a Change in Diesel Emission Standards

Assumptions about car buyer's response to imposition of a 0.2g/mi standard	Assumptions about impacts of dieselization		
	Maximize benefit of shift to 0.2g/mi	Intermediate	Maximize benefit of maintaining 0.6g/mi
(1) No change in diesel share of 25% of sales	A1	B1	C1
(2) Diesel share holds at 10% of sales; former diesel owners switch to gasoline-powered vehicles of same fuel economy but lighter weight, smaller size	A2	B2	C2
(3) Diesel share holds at 10% of sales; former diesel owners switch to gasoline-powered vehicles of same weight but lower fuel economy	A3	B3	C3

- **Social Cost of Increased Oil Use** that is increased when
  consumers shift from diesels to less efficient gasoline
  vehicles of the same size in response to increased costs of
  diesel particulate control systems;
- **User Costs** that are imposed when car buyers who want diesels
  must pay more for diesel vehicles or purchase less desirable
  vehicles.

The magnitude of uncertainties has led the committee to omit
several dimensions of cost and benefits in the analysis, and to deal
with others in summary ways. As already indicated, these include
pulmonary and other human systemic disease, environmental effects other
than visibility, and non-fatal accident injuries.

Sometimes environment-related cost-benefit studies underestimate
hard-to-quantify environmental and health benefits. To assure that the
best case is made for the shift to the 0.2 g/mi standard for diesel
particulates, a number of assumptions have been made in all three cases
that serve to increase the benefits of the shift. The principal
assumptions are that (1) all vehicles in the light-duty diesel auto and
truck fleet emit particulates at 0.6 g/mi (though some small vehicles
in the fleet already emit below this level); (2) gasoline vehicles that
replace diesels, when particulate control costs diminish diesel sales
to 10 percent, are entirely free of diesel-like particulate emissions
(though at least one study indicates that gasoline vehicles also emit a
small quantity of carcinogenic particulates); and (3) rural populations
are assumed to be exposed to low, but significant, concentrations of
diesel particulate emissions.

Significant problems also occur in comparing those effects that
have been explicitly quantified in the analysis. Even similar appearing
effects are difficult to compare, such as the person-years of life
saved by reducing the cancer risk and the person-years of life lost by
the diminished protection against road accidents afforded by a lightweight fleet of autos. The comparison is made in terms of person-years
of life extended (or shortened) rather than lives lost to take account
of the age difference between traffic and cancer victims. Mortality
statistics indicate that traffic victims are younger on the average
than the population at large, while cancer victims are substantially
older. Based on recent mortality statistics, a death delayed by
averting a fatal road accident is considered to affect longevity by a
factor of 2.6 greater than a death delayed by averting a case of lung
cancer (U.S. Department of Commerce, 1979; U.S. Department of Health,
Education and Welfare, 1980). While this effect is taken into account
in Table 7.4, the committee has made no allowance for the possible
differences in patterns of morbidity associated with the various health
and safety effects. Moreover, the estimates shown reflect a simple sum
of the total effects on cancer incidence and only the annual effects on
road accidents, without allowing for the cancer effects that will
appear over a lengthy period.

TABLE 7.3 Numerical Values of Assumptions in Analyzing Cases for Diesel Sales of 25 Percent in 1995

	Case Set A: Maximum Benefits of 0.2 g/mi			Case Set B: Intermediate			Case Set C: Maximum Benefits of 0.6 g/mi		
	A1	A2	A3	B1	B2	B3	C1	C2	C3
a. Diesel Vehicles in Use (millions post-1994)	29.5	11.6	11.6	29.5	11.6	11.6	29.5	11.6	11.6
b. New Fleet Fuel Efficiency (mpg)	27.5	27.5	26.8	27.5	27.5	26.8	27.5	27.5	26.8
c. Reduction in Weight: (pounds per vehicle)	0	75	0	0	75	0	0	75	0
d. Particulates Removed (10^3 metric tons)	140	180	180	140	180	180	140	180	180
e. Particulate Impacts (person-years saved per 10^3 tons particulate removed)	255	255	255	17	17	17	0	0	0
f. Visibility Impacts (dollar benefits per ton particulate removed)	2022	2022	2022	926	926	926	0	0	0
g. Traffic Safety Impacts (person-years lost per pound of vehicle weight reduction)	150	150	150	300	300	300	600	600	600
h. Oil Import Reduction Premium (dollars per gallon)	.10	.10	.10	.25	.25	.25	.50	.50	.50
i. Fleetwide Fuel Savings (10^6 gallons per year)	0	0	1700	0	0	1700	0	0	1700
j. 0.2 g/mi Standard Control Costs (dollars per vehicle per year)	19	19	19	52	52	52	78	78	78

Discussion of the basis for these assumptions can be found on the following pages in this report and the references cited there: a. pp. 87-89, 105-106; b. pp. 8-10, 82-84; c. pp. 84-86, 115-116; d. pp.

TABLE 7.4 Benefits and Costs in 1995 from Changing Particulate Standard from 0.6 g/mi to 0.2 g/mi

	Cases Most Favorable to 0.2 g/mi Set A			Intermediate Cases Set B			Most Favorable to 0.6 g/mi Set C		
	A1	A2	A3	B1	B2	B3	C1	C2	C3
Benefits:									
A. Cancer Reduction (Person-years saved)	36,000	46,000	46,000	2,400	3,100	3,100	0	0	0
B. Visibility Improvement (10^6 dollars)	280	360	360	130	170	170	0	0	0
Costs:									
C. Safety Reduction (Person-years lost)	0	11,000	0	0	22,000	0	0	45,000	0
D. Import Reduction Premium (10^6 dollars)	0	0	170	0	0	430	0	0	850
E. Increased User Costs (10^6 dollars)	560	390	390	1,500	1,040	1,040	2,250	1,550	1,550
Summary:									
F. Person-Years Saved (Lost)	36,000	35,000	46,000	2,400	(18,900)	3,100	0	(45,000)	0
G. Resource Gain (Loss) (10^6 dollars)	(280)	(30)	(200)	(1,370)	(870)	(1,300)	(2,250)	(1,550)	(2,400)
H. Resource Loss Per Person-Year Saved (in dollars)	7,800	860	4,300	571,000	--	419,000	--	--	--

In most cases these values can be derived straightforwardly from the assumptions of Table 7.3 (with adjustments for rounding). Examples drawn from Case A illustrate the method. The way of deriving the Summary figures is also given.

$$F = A - C$$
$$G = B - D - E$$
$$H = -G/F$$

A = d x e   36,000 = 140 x 255
B = d x f   280 = 140 x 2022
C = c x g   11,000 = 75 x 150
D = h x i   170 = .10 x 1700
E = a x j   560 = 29.5 x 19

Comparison of Costs and Benefits

Table 7.4 illustrates in disaggregated form the outcomes in each of the nine alternative cases. If the net benefits of shifting to 0.2 g/mi were consistently positive or negative in all cases, it would be possible to reach a conclusion without further attempts to compare the various impacts. In fact, Table 7.4 shows this is not the case.

Case C1 presents results for the set of assumptions minimizing the net benefits of shifting to 0.2 g/mi and assuming no reduction in diesel sales because of emission control costs. In this case the shift to a 0.2 g/mi standard saves no person-years in 1995 and costs $2,250 million in increased user costs. The numbers indicate a negative net benefit, so the 0.2 g/mi standard would not be desirable in Case C1. The results in Cases C2 and C3 are even less favorable to the 0.2 g/mi standard. The rejection of the 0.2 g/mi standard under assumptions favorable to the 0.6 g/mi standard is hardly conclusive however.

To obviate the need for further information, it would be necessary to show that the 0.6 g/mi standard was preferred under all of the nine cases. Do the outcomes in the other cases indicate that the 0.6 g/mi standard is a dominant alternative?

Case A can be used to test for dominance of the 0.6 g/mi standard, for it incorporates a set of assumptions that maximize the benefits of the 0.2 g/mi standard. In Case A1, with no change in the diesel market, the 0.2 g/mi standard saves 36,000 person years annually and provides $280 million worth of visibility benefits at an increased user cost of $560 million. Subtracting costs from benefits yields a net cost of $280 million for saving 36,000 person years. To evaluate this outcome it is necessary to compare the avoidance of premature deaths with the additional costs imposed on diesel buyers. If society assesses the worth of an additional year of life to one individual at more than $7,800, the 0.2 g/mi standard is preferred under the assumptions of Case A1; if less, then the 0.6 g/mi standard is preferred in Case A1.

Case B1 presents an intermediate situation with a net resource loss of $1,370 million and a savings of 2,400 person years. In the three B cases, the net benefits or costs of the shift to 0.2 g/mi depend on the uncertainties so critically that a conclusion cannot be drawn without explicitly setting a value on a "person-year saved." The ratio of resource loss to person-years saved indicates the value of person-years saved above which Case B1 would indicate a net positive benefit to the shift to 0.2 g/mi and below which there is net cost to the shift in standards. In order to understand whether this ratio of $571,000 per person-year saved is higher or lower than the value of a person-year saved, the average cost per person-year extended can be compared to other expenditures, social or private, for prolonging life for terminally ill patients or in averting risks of death.

Integration of Costs and Benefits

Table 7.5 presents data on the "value of life" implicit in government programs and in several studies of occupational risks and automobile use compiled by Bailey (1980). The wide spread for the government programs does not provide a good benchmark for evaluating the implied costs of life-saving in Case B for two primary reasons: Within the government decision processes there are means to limit program budgets but not to assure adequate and consistent levels of program cost-effectiveness. This can and does result in both direct assistance programs (e.g., health care) and regulatory programs (e.g., coke oven standards) with a wide range of cost-effectiveness--in this case measured by the value of person-years saved. The high levels of the value of person-years saved implicit in federal programs are clearly unrealistic guidelines for the majority of federal actions, because they are not affordable. If implemented across the board, there would not be enough gross national product to cover all the public and private expenditures.

Assuming that willingness-to-pay estimates of the life values are typical of death delays equal to 20 person years, then the value of a person-year prolonged ranges from $8,000 to $35,000.* The implied value of a person-year saved in Case A1 is at the lower end of the range of the evaluations. The values in Case B1 are clearly above the high end of the range.

If the stricter standards cause diesel sales to level off at 10 percent instead of 25 percent, the 15 percent of vehicle users who prefer to own diesels, but are priced out of the diesel market by the higher control costs, may shift to gasoline vehicles of the same fuel efficiency (Case A2), or they may shift to gasoline vehicles of equal weight to the displaced diesels (Case A3) or to any intermediate case. Cases A2 and A3 have lower costs and greater benefits compared with Case A1. Case A2 involves an increase in traffic deaths, but no change in fuel consumption, while Case A3 results in an increase in oil consumption, but no change in road deaths. In Case A2, the net result is that the 0.2 g/mi standard produces an average cost of about $860 per person-year saved, which is a value well below other implicit or explicit expenditures made on life-prolonging activities. Under the assumption of Case A3 that the shift from diesels is to gasoline-powered cars of equal weight, the average cost per person-year extended is about $4,300, a value also below the range implied by Table 7.5.

In summary, under the assumptions of the Case A set favorable to the 0.2 g/mi standard, and given the commensurability of environmental effects with other dollar magnitudes, the choice of the stricter standard implies that the value per person-year of life extended may

---

*The actual number of person-years saved per death delayed is a function of the cause of death, and the willingness-to-pay value of person years may also be a complicated function of age at premature death as well as the cause of death. Use of the 20 person-year concept is a simplification.

lie in the range from $860 to $7,800. This range is considerably below the values designated by both private behavior and public policies implied in Table 7.5.

Applying the same kind of analysis to the Case B set, which has intermediate assumptions about health effects and control costs, produces a cost per person-year saved of $419,000 or more for the 0.2 g/mi standard. This cost is well above those typically encountered in federal programs or contexts and indicates that a standard above 0.2 g/mi would be preferred if the assumptions of Case B were true.

Policy Making Under Uncertainty

The cost-benefit analysis performed in the study indicates that the 0.2 g/mi standard would be consistent with other programs and policies in situations in which the assumptions of the Case A set were true, but that a standard above 0.2 g/mi would be indicated under conditions encountered in cases B or C. If more were known about the probability of cases A, B, or C, certain other techniques of decision theory could be used to develop expected values for different outcomes. Unfortunately, in assessing the available knowledge of the relevant factors, the committee found it impossible to make quantitative estimates of the joint probability distribution of the many elements that enter the evaluation of the alternative regulatory standards. The committee was unable to assign probabilities to the intermediate situations in the Case B set. The committee holds, however, that case sets A and C represent plausible extremes for assumptions favorable to the 0.2 g/mi and the 0.6 g/mi standards, respectively. The subsequent discussion assumes that these are reasonable representations of extreme cases for the purpose of formulating public policy.

When decisions must be made under conditions of great uncertainty, it is not sufficient to consider only the outcome if uncertain variables assume expected or predicted values. It is important to take the results into account also if uncertainties are resolved in ways adverse to the alternative selected. This involves examining the consequences if we choose the 0.2 g/mi standard and uncertain variables take the values assumed in Case C or, alternatively, if we choose the 0.6 g/mi standard and the assumptions of Case A are actually realized. Such an analysis permits us to estimate:

- The loss under each of the regulatory alternatives, if the uncertainties are resolved in a way adverse to the alternative chosen. Thus, as discussed earlier, the outcome, if the assumptions of Case C prove to be correct and we have chosen the 0.2 g/mi standard, measures the loss or "regrets" for the choice of that standard. The "regrets" are therefore a measure of the risk imposed by uncertainty.
- Additional information may reduce the uncertainty associated with a policy choice, thus reducing the magnitude of the "regrets." The reduction is an index of the value of such information.

TABLE 7.5   Value of Life Implied by Federal Regulations and Individual Behavior

	Cost per Life Saved (millions of dollars)
Federal Regulation	
Coke-oven emission standard (OSHA)	4.5 to 158
Lawn mower safety standard (CPSC)	0.2 to 1.9
Occupational exposure to acrylonitril standard (OSHA)	1.9 to 624.9
Willingness to Pay(1978 dollars)	
Occupational safety (1977-1978)	0.170 to 0.584
Seatbelt users (1979)	0.256 to 0.715
Occupational safety (1979)	0.376

OSHA – Occupational Safety and Health Administration
CPSC – Consumer Product Safety Commission

Source: Data obtained from sources cited in Martin J. Bailey, Reducing Risks to Life: Measurements of the Benefits, Washington, D.C.: American Enterprise Institute, 1980.

To illustrate the analysis of regrets, consider the regrets associated with the choice of the 0.6 g/mi standard. As reflected by the Case A assumption, society loses the opportunity to save between 35,000 and 46,000 person-years per year at a cost of $30 million to $280 million per year. The magnitude of regrets depends on the value for a person-year saved. While the committee's Analytic Panel has reached no judgment on this, it has calculated the regrets under two alternative assumptions about the value of a person-year saved. At a value of $10,000 per person year saved--a value chosen from the low end of the range of willingness-to-pay estimates in Table 7.5--the net loss of accepting the particulate standard of 0.6 g/mi, given the assumptions of Case A, would be between $80 million and $320 million per year. In the opposite situation, under the assumption of Case C, the 0.2 g/mi standard results in the lose of as many as 45,000 person-years and between $1.6 billion and $2.4 billion per year in economic costs. If a person-year saved is valued at $10,000, the net loss ranges from $2 billion to $2.4 billion per year (see Table 7.6).

At a value of $30,000 per person-year saved, the net loss of choosing the 0.6 g/mi standard, given the assumptions of Case A, would be between $800 million and $1.18 billion. In the opposite situation, if the 0.2 g/mi standard is chosen under the assumptions of Case C, the regrets range from $2.25 billion to $2.9 billion.

The above analysis of regrets suggests that risks associated with the adoption of the 0.2 g/mi standard are greater than those for the adoption of the 0.6 g/mi standard. At a value per person-year of $10,000, the maximum regrets associated with the 0.2 g/mi standard are several times greater than those for the 0.6 g/mi standard. At the higher value per person-year of $30,000, the maximum regrets for the 0.2 g/mi standard are more than double those for the 0.6 g/mi standard. Only at an extremely high value per person-year (exceeding $1.75 million) would the maximum regrets associated with the adoption of the 0.6 g/mi standard exceed those for the 0.2 g/mi standard.

In either case, the costs of making an incorrect decision would clearly pay many times over for the necessary research to reduce the range of uncertainty and lead to better decision making. The analysis of regrets here reaches conclusions similar to a more detailed analysis performed by Hogan (1981).

Table 7.6 Regrets Associated with Policy Choices (in millions of dollars)

	Value of Person-Year	
	$10,000	$30,000
Choice of 0.6 g/mi	80--320	800--1180
Choice of 0.2 g/mi	2000--2400	2250--2900

An Alternative Policy Analysis

Of necessity, the few cases examined by the committee overlook attractive options for the decision maker and draw too sharply the extreme differences among alternative futures. For instance, a definitive judgment on the nature of pollution control technology and the standard for particulate emissions cannot be made. At best the administrator of the EPA can hope to control these variables for a few years only; after that, a new Congress or new EPA administrator may revise the laws and reevaluate the standards. Future choices will be made with the advantage of more market experience, new and improved health research data, better understanding of environmental processes, and further advances in control technology--a wider context of information that is likely to be critical to better assessment of the balance between transportation, health, the environment, and the common weal.

Knowing there is time and opportunity to collect and examine the information in order to resolve critical uncertainties about demand, costs, health, and environmental effects, and then to make decisions based on the best information available, could affect the choices available today. If there is a low risk of serious health damage from particulates, the growth of a small-scale fleet of diesel cars could be permitted while research continued on health and environmental effects. Should the health and environmental risks later be proven, the expansion of the diesel fleet could be halted. Hence, even if the probability of health risks is serious enough to preclude an immediate commitment to a large diesel fleet, there may be opportunities to experiment with smaller numbers of diesel passenger cars pending the outcome of additional research on health effects.

However, the foregoing analysis is silent on such sequential strategies. Analyzing sequential regulatory strategies is both a difficult conceptual problem and a complex calculational effort. Notwithstanding, the "decision analysis" approach attempts to examine such strategies.

The details of one expanded decision analysis appear in Hogan's paper (1981) prepared for the committee. The results confirm the general insights of the committee's analysis discussed above. The Hogan analysis shows that the decision about the appropriate choice of a diesel particulate standard, 0.6 g/mi versus 0.2 g/mi, depends critically upon the outcome of such uncertain situations as the level of demand, the severity of health effects, the cost of control technology, the importance of visibility. No single standard is best in all cases. Furthermore, in all but the most extreme circumstances there is no advantage to making an early decision impose to the more stringent standard. Hogan concludes that development efforts should be directed at lowering the cost of effective particulate control technology. Testing should proceed to narrow the uncertainty regarding potentially adverse health effects. While development and testing are proceeding, more data will become available about the strength of the market for diesels. When all this comes together, analysts will be able to review the standard more fully and reach a better judgment about the best level of emissions control for diesel cars and small trucks in model years in the late 1980's.

Hogan finds that even in the case where the lower standard is more appropriate, the relatively low level of diesel market penetration that would occur ensures that the aggregate damages of the delay will be small. Assuming the high costs of an early imposition of the more stringent standard, the low probability of serious health hazards, and the chance to make course corrections in the standard, the preferred decision in Hogan's analysis is to retain the higher 0.6 g/mi standard with the option of tightening it later.

ALTERNATIVE APPROACHES TO REGULATING DIESEL PARTICULATES

Current EPA regulations implement a two-step strategy to realize two different objectives: (1) the development of new options for controlling diesel particulate emissions and (2) the technological limitation of emissions to the lowest level achievable by the best available control technology. To provide an incentive for manufacturers of emissions control systems and diesel passenger cars and light trucks, the EPA states, in effect, that the reward for successful research and development will be a market for control technologies that limit diesel particulate emissions to 0.2 g/mi. In its second objective, EPA is required to promote a balance among several factors: low emissions, availability, cost, noise, energy, safety, and lead-time.

EPA's current approach to achieving those objectives, by its adherence to a stringent standard of 0.2 g/mi for vehicles sold in 1985 and thereafter, may lead to costly mistakes, as suggested in the above analysis of regrets. To avoid the possibly high costs of incorrect decisions, the committee concludes that the EPA should retain the 0.6 g/mi standard for post-1984 vehicles and commit to a formal reevaluation of its regulatory decision on the basis of the improved information about the current uncertain health, environmental, technical and economic factors that affect the outcome of its regulatory decisions. Such an approach is particularly suited to the problem of regulating diesel vehicle emissions both because the uncertainties are currently great and because the introduction of diesel vehicles into the light-duty vehicle fleet will proceed gradually. As a consequence the risks and benefits of regulatory policies will accrue slowly, and there will be ample opportunity to revise regulatory decisions on the basis of emerging information if the parties to the decision continue to address the issue. A commitment by the EPA to formally reevaluate the need for more stringent standards can contribute to the incentives of manufacturers to continue to gather information and to develop appropriate control technology. Such reevaluations should be conducted at appropriate intervals as determined by new information. This approach is consistent with a sequential decision-making process best suited to situations laden with uncertainties.

- Regulate other diesel emission sources in road transport, such as heavy trucks and city buses;
- Explore intermediate levels of diesel particulate emission standards between 0.6 g/mi and 0.2 g/mi;
- Control appropriately defined total emissions for the whole diesel fleet instead of individual vehicles;
- Apply state standards rather than nationwide ones; and
- Levy emissions charges in lieu of emission standards.

These alternatives are described and compared on the next three pages.

Regulate Other Diesel Emission Sources First

Despite high sales projections, diesel cars and small trucks are likely to increase their percentage in the fleet at a slow rate. Projections of light diesel sales of 25 percent and 50 percent may be optimistic. Light-duty vehicles have a life expectancy of about ten years. Thus, many years will be required for the fleet to reach a stable level of diesel vehicles. For example, the diesel share of new car sales could rise linearly to 25 percent in 1995, but in 1995 only 14 percent of the fleet would consist of diesels. Diesels would not constitute 25 percent of the fleet until 2005. Since the characteristics of the stock of total vehicles lags behind new car sales, it would be possible to monitor the fleet composition and to modify standards only if the number of diesel passenger cars and light trucks becomes significant in actual use.

If the ambient level of diesel-like particulates is to be maintained at the 1980 levels or lowered, it may be efficient and sufficient to regulate such alternative sources of particulates as heavy-duty diesels before imposing stricter standards than 0.6 g/mi on light-duty vehicles. Even at 0.6 g/mi beyond 1985, the annual aggregate particulate emissions in light-duty diesel in 1995 would still be less than the annual particulate emissions from large diesel vehicles in 1980. A heavy-duty diesel vehicle currently emits roughly 30 times more particulates over its lifetime than a light-duty diesel that does not meet the 0.6 g/mi standard. This suggests that accelerating the control of heavy-duty diesel emissions should be a high priority for EPA.

Limiting emissions from heavy-duty diesels may be significantly more efficient than controlling emissions from light-duty diesels. Based on EPA's regulatory analysis (1980), the Analytic Panel estimates that the 0.2 g/mi standard for light-duty diesels is likely to cost between $2,500 and $3,150 to reduce tailpipe emissions by 1 metric ton of diesel particulate. Heavy-duty diesels emit so much more particulate matter that the control systems for a single vehicle would reduce emissions by roughly 1 metric ton each year. Because it is difficult to imagine even a super-effective control system for large diesels that would cost as much as $2,500 per year, it is likely that heavy diesels would be significantly more cost-effective than control systems for diesel cars.

Explore Intermediate Emission Standards

Imposing a diesel average particulate standard for at least a short period, at a level such as 0.4 g/mi for six model years, would probably generate a substantial amount of information about the cost and effectiveness of different control systems on different types and sizes of vehicles. This information could contribute greatly to synthesizing a final uniform standard for all diesel cars and light trucks in the late 1980's. An interim approach to a diesel exhaust particulate standard could be imposed sooner, with greater certainty than the 0.2 g/mi standard to achieve the same or lower aggregate fleet emissions. For example, in the case where the diesel car market equals 25 percent, an average particulate standard of 0.4 g/mi imposed from 1985 through 1990 creates about the same six-year aggregate emissions as the imposition of a 0.2 g/mi standard in model year 1988.

Control of Total Diesel Fleet Emissions

Diesel particulate emission rates vary by engine size. Small diesel engines currently produce fewer particulate emissions than large diesel engines. With engine modifications, small light-duty diesels can approach and even surpass the 0.2 g/mi standard without trap-oxidizers, while large light-duty diesels cannot. Because the cost of reducing particulate emissions is likely to vary with engine size, it could be less costly to users to adopt a diesel fleet average emissions standard than to adopt a uniform emission standard. A Diesel Average Particulate Standard would allow manufacturers to meet the standard in the aggregate, while allocating the cost of particulate reduction more equally across engine sizes. Compared to uniform standards, this approach would probably decrease control costs and still meet the standards on the average. This alternative meets some but not all of the objections to a uniform standard. Although promising efficiency gains, this approach may have certain drawbacks in terms of equity. Vehicles of the same size made by different manufacturers could have different emission rates. Moreover, the sales mix within firms could affect their differential control costs more than is presently the case.

State Standards

Currently there is a single national emissions standard, though California is allowed to set more stringent standards. Air pollution problems tend to be confined to the major metropolitan areas; most rural areas do not have serious air pollution problems. Accordingly, following the principal of the separate California standard, an alternative to national standards would enable states to establish standards for vehicles operated in their areas. This strategy would be appreciably less costly than the current one because expensive emissions controls would not be required where they are not needed.

Whether there would be enforcement problems, with migrant vehicles confounding the local emission limit, would depend on the nature and scope of the standard. If it resulted in many states imposing different emission standards, the costs of vehicle production and regulatory administration and enforcement would be increased.

However, some types of local regulations probably would not incur such costs--e.g., banning diesel taxicabs from New York City streets. Another option is local license fees based on EPA certification of emission rates.

Emission Charges

Comparisons of the uses and limits of emission charges with emission standards indicate that fees are more appropriate and effective than mandatory standards in various circumstances (Spence and Weitzman, 1977). The total sacrifice by society for limiting pollution to a predetermined level can be lower with the use of emission charges than mandatory standards if the costs of reducing pollution vary among automobile owners and vehicle types.

Emission charges also keep a steady pressure on manufacturers to reduce pollution below the standard and automatically incorporate current cost considerations into their plans. For example, a manufacturer might develop a low cost technology slowly, starting out above the standard and perfecting it over time. The steady financial pressure caused by an emission fee can also produce desirable intermediate outcomes. A manufacturer might be able to meet a 0.3 g/mi level, but not a 0.2 g/mi level. An emission fee would automatically encourage a manufacturer to do this, whereas a standard regulation might just lead to further delay of the implementation of the 0.2 g/mi level. The potential advantages of emission fees should be assessed, together with an analysis of their effects on various parts of society and the requirements for their implementation and administration.

COMPARISON OF ALTERNATIVE REGULATORY APPROACHES

Two characteristics of the regulatory approaches are important: their efficiency and their equity.

An efficient regulatory approach is one that is least costly at each level of emission reduction. A regulation's efficiency can be closely linked to its accompanying incentive structure. Since the prospective 1985 particulate emission standard is not considered feasible with proven pollution control equipment, it is, in effect, "technology forcing"--i.e., it is intended to apply pressure on manufacturers to develop new methods of emission control.

Both the average diesel exhaust particulate standard and alternative source regulation offer improvement in efficiency over the uniform approach. If diesel emissions turn out to be a problem in only a few urban areas, local regulation is likely to be more efficient.

Different forms of regulations may treat firms or consumers unequally. For instance, current national standards for emissions impose costs on consumers in areas with no air quality problems and provide few benefits. A regulatory alternative that allows sales-weighted averaging of emissions, such as diesel exhaust particulate standards, may provide advantages to firms that produce full product lines. Enforcement of complex regulatory schemes, such as local controls might incur, may be particularly onerous for small firms. For example, certification procedures may impose large per unit costs on small volume producers.

In developing its regulatory approach, EPA was confronted with the alternatives described here, as well as several others. Its reasons for rejecting them fall into two categories:

* The expected costs, described in terms of enforcement and "side-effects," would outweigh the increased benefits of regulatory efficiency, and
* The alternative regulatory approach would violate existing statutes or the EPA interpretation.

The statutory requirements can be interpreted as forbidding similar vehicles from emitting pollutants at different levels, thereby ruling out a diesel average particulate standard or forcing EPA to require all sources to use the best available control technology, as opposed to providing incentives for the industry to reach a least overall cost solution by equilibrating marginal control costs. If EPA is interpreting the legislative requirements properly and if the arguments are at least convincing enough to suggest that EPA be allowed to seriously consider alternative strategies, then there is a strong reason to support amending the Clean Air Act in order to further serious attention to the alternative approaches.

REGULATORY INTERVENTION

In recent years a vast literature on regulation has appeared. The circumstances in which regulatory programs were established have been chronicled. The political and institutional dynamics of regulation have been described. The legal procedures and policies of regulatory systems have been examined. The economic performance of regulated industries and the costs and benefits associated with regulation have been weighed, measured, and analyzed. Still, little of such analyses bears on the question that policy makers most need to answer: What are the likely effects of imposing and implementing a regulatory strategy or, alternatively, of not going ahead with certain regulations and standards?

Chapter 6 spoke of "market failure" that may justify a regulatory intervention and "externalities" that impose costs of activities on third parties—that is, on people who do not fully benefit from them (or _vice versa_). But if market conditions are almost always flawed, so are regulatory interventions by government. However imperfect consumer information may be about such matters as performance, safety, and durability of a product, this pales by comparison to what a regulatory agency needs to know in prescribing standards for worker safety, say, or making sure that nuclear power plants have adequate safeguards or requiring pollution controls on industrial firms to protect or improve the welfare of millions of people while taking into account the dynamic technical and economic realities of tens or hundreds of firms. The actions taken by regulatory agencies also beget externalities. While market transactions of unsafe products will often put third parties in harm's way, regulatory rules sometimes cause unforeseen and unintended risks for consumers, workers, stockholders, and other third parties. This could be called "regulatory failure." The general problem of failures in public policies that are designed to remedy instances of market failure has been considered by Wolf (1979).

Cost-benefit analyses can be helpful in coming to grips with regulatory failure, though such exercises are also invariably flawed and their guidance for policy limited. The many reasons for this have been discussed: the difficulty of identifying and quantifying costs and benefits; the incommensurable nature of valuations of human life or health; the special difficulty of evaluating extremely low risk but serious problems, especially when the hazard is complicated by external events and personal behavior; the problem of interpersonal and intergenerational comparisons of effects and utility; the necessity of making (or accepting) trade-offs; the uncertainty or lack of information about the consequences of taking an action or taking any alternative courses of action. Cost-benefit analyses, therefore, often result in clearer understanding of the options, but they do not provide definitive answers. As a technique in the hands of decision-makers who are confronted regularly with the necessity of making choices that involve setting standards, issuing regulations, and so forth, cost-benefit analyses are useful. They can supplement but not substitute for the exercise of informed judgment.

REFERENCES

Bailey, Martin J. (1980). <u>Reducing Risks to Life: Measurement of the Benefits</u>. Washington, D.C.: American Enterprise Institute.

Brookshire, David S., Ralph D. d'Arge, William D. Schulze, and Mark A. Thayer (1979). <u>Methods Development for Assessing Air Pollution Control Benefits. Vol. II.</u> Experiments in Valuing Non-Market Goods: A Case Study of Alternative Benefit Measures of Air Pollution Control in the South Coast Air Basin of Southern California. EPA 600/5-79/001b. Washington, D.C.: U.S. Environmental Protection Agency, Office of Health and Ecological Effects.

Harris, Jeffrey E. (1981). <u>Potential Risk of Lung Cancer from Diesel Engine Emissions</u>. Report to the Diesel Impacts Study Committee. National Research Council. Washington, D.C.: National Academy Press.

Hogan, William W. (1981). Decision Analysis of Regulating Diesel Cars. Report to the Diesel Impacts Study Committee, National Research Council. Washington, D.C.: National Academy Press.

National Research Council (1979). <u>Airborne Particles</u>. Report of the Committee on Medical and Biological Effects of Environmental Pollutants. Washington, D.C.: National Academy of Sciences.

National Research Council (1981a). <u>Health Effects of Exposure to Diesel Exhaust</u>. Report of the Health Effects Panel of the Diesel Impacts Study Committee. Washington, D.C.: National Academy Press.

National Research Council (1981b). <u>Diesel Technology</u>. Report of the Technology Panel of the Diesel Impacts Study Committee. Washington, D.C.: National Academy Press (forthcoming).

Rowe, Robert D., Ralph d'Arge, and David S. Brookshire (1980). An Experiment on the Economic Value of Visibility. <u>Journal of Environmental Economics and Management</u>. 7(1):1-19.

Spence, A. M., and M. L. Weitzman (1977). "Regulatory Strategies for Pollution Control." In <u>Approaches to Controlling Air Pollution</u>, A. F. Friedlander, ed. Cambridge, MA: The MIT Press, pp. 199-219.

U.S. Department of Commerce. Bureau of the Census (1980). <u>Statistical Abstracts of the United States</u> (101st edition). Washington, D.C.

U.S. Department of Health, Education and Welfare (1980). <u>Health, United States</u>. Public Health Service, Office of Health Research, Statistics and Technology, National Center for Health Statistics. Hyattsville, MD. Publication No. 80-1232.

U.S. Environmental Protection Agency (1980a). New Emission Estimates for Highway Vehicles. Memorandum from Charles L. Gray, Jr., Emission Control Technology Division, Ann Arbor, MI.

U.S. Environmental Protection Agency (1980b) Air Quality Criteria for Particulate Matter and Sulfur Oxides. Vol. III. Ch. 10.

U.S. Environmental Protection Agency (1980c). Regulatory Analysis: Light-Duty Diesel Particulate Regulations. February 1980.

Wolf, Charles, Jr. (1979). A Theory of Non-Market Failures. <u>The Public Interest</u>. Spring 1979:114-133.

APPENDIX A

Panels, Consultants, and Contributors

ANALYTIC PANEL

Fred S. Hoffman, <u>Chairman</u>
Senior Economist
Rand Corporation

Sandford F. Borins
Assistant Professor of Policy and Environment
York University
Downsview, Ontario, Canada

William M. Capron
Professor of Economics
Boston University

Donald N. Dewees
Associate Professor of Economics
Institute for Policy Analysis
University of Toronto

Jeffrey E. Harris
Associate Professor of Economics
Massachusetts Institute of Technology

William W. Hogan
Professor of Political Economy
John F. Kennedy School of Government
Harvard University

Gregory K. Ingram
Senior Economist
The World Bank

James P. Wallace, III
Vice President and Division Manager
The Chase Manhattan Bank

ENVIRONMENTAL IMPACTS PANEL

Sheldon K. Friedlander, _Chairman_
Vice-Chairman
Department of Chemical, Nuclear, and Thermal Engineering
University of California at Los Angeles

Alan Q. Eschenroeder
Senior Staff Scientist
Arthur D. Little, Incorporated

Eville Gorham
Professor of Botany and Ecology
University of Minnesota

Jack D. Hackney
Professor of Medicine
Rancho Los Amigos Hospital (University of Southern California)

Ronald A. Hites
Professor of Public and Environmental Affairs
Indiana University

Issac R. Kaplan
Professor of Geology and Geochemistry
University of California at Los Angeles

David B. Kittelson
Professor of Mechanical Engineering
University of Minnesota

Arthur M. Winer
Assistant Director
Statewide Air Pollution Research Center
University of California at Riverside

HEALTH EFFECTS PANEL

Herschel E. Griffin, <u>Chairman</u>*
Associate Director and Professor of Epidemiology
Graduate School of Public Health
San Diego State University

David J. Brusick
Director of the Department of Genetics and Cell Biology
Litton Bionetics, Incorporated

Neal Castagnoli, Jr.
Professor of Chemistry and Pharmaceutical Chemistry
University of California at San Francisco

Kenneth S. Crump
President
Science Research Systems, Incorporated

Bernard M. Goldschmidt
Associate Professor of Environmental Medicine
New York University Medical Center

Ian T. Higgins
Professor of Epidemiology
University of Michigan

Dietrich Hoffman
Associate Director of the Naylor Dana Institute for Disease Prevention
American Health Foundation

Steven M. Horvath
Director of the Institute of Environmental Stress
University of California at Santa Barbara

*Former Dean of the Graduate School of Public Health
 University of Pittsburgh

Paul Nettescheim
Chief of the Laboratory of Pulmonary Function and Toxicology
National Institute of Environmental Health and Safety

Bruce O. Stuart
Manager of the Inhalation Toxicology Section
Stauffer Chemical Company
and Clinical Associate
University of Connecticut Medical School

Hanspeter Witschi
Senior Research Staff Member, Biology Division
Oak Ridge National Laboratory

TECHNOLOGY PANEL

Robert F. Sawyer, <u>Chairperson</u>
Professor of Mechanical Engineering
University of California at Berkeley

Frederick L. Dryer
Associate Professor of Mechanical Engineering
Princeton University

James R. Johnson
Executive Scientist (Retired)
3M Company

James R. Kliegel
President
KVB, Incorporated

Paul Kotin
Senior Vice President for Health, Safety and Environment
Johns-Manville Corporation

William J. Lux
Director of the Product Engineering Center
John Deere Industrial Equipment Division
Deere & Company

Phillip S. Myers
Professor of Mechanical Engineering
University of Wisconsin

James Wei
Professor of Chemical Engineering
Massachusetts Institute of Technology

CONSULTANTS AND CONTRIBUTORS

Jack Appleman
Vice President
Lewin and Associates, Incorporated

Alfred G. Catteneo
Consulting Engineer
Berkeley, California

Nicholas P. Cernansky
Associate Professor of Mechanical Engineering
Drexel University

David Christison
Manager of Planning, Technology and Financial Analysis for Manufacturing
    (retired)
Mobil Oil Corporation

Renee G. Ford
Consulting Editor
Rye, New York

Daniel Grosjean
Technical Director, Applied Research Operations
Environmental Research & Technology, Incorporated

Susanne V. Hering
Research Engineer, Department of Chemical Engineering
University of California at Los Angeles

John H. Johnson
Professor of Mechanical Engineering
Michigan Technology University

Lloyd Johnson
Leijon Engineering Company
East Peoria, Illinois

Frank Marzulli
Health Consultant
Bethesda, Maryland

Ronald D. Matthews
Associate Professor of Mechanical Engineering
University of Texas at Austin

Roy McDonald
Graduate Student
Stanford Graduate School of Business

Sharon Rasmussen
Consulting Editor
Washington, D.C.

Owen I. Smith
Assistant Professor of Engineering
University of California at Los Angeles

Alan P. Waggoner
Associate Profesor of Engineering
University of Washington

Lawrence J. White
Professor of Economics
New York University Graduate School of Business

Dennis Yao
Consultant
Palo Alto, California

PUBLICATIONS OF THE STUDY

Committee

National Research Council. Critique of the U.S. Environmental Protection Agency's <u>Initial Review of Potential Carcinogenic Impact of Diesel Engine Exhaust</u>. Letter to David G. Hawkins, Assistant Administrator For Air, Noise, and Radiation, EPA, from Henry S. Rowen, Chairman, Diesel Impacts Study Committee. December 6, 1979.

Panels

National Research Council. <u>Health Effects of Exposure to Diesel Exhaust</u>. Report of the Health Effects Panel, Diesel Impacts Study Committee. Washngton, D.C.: National Academy Press, 1981.

National Research Council. <u>Diesel Technology</u>. Report of the Technology Panel, Diesel Impacts Study Committee. Washington, D.C.: National Academy Press, forthcoming.

Supporting Papers

Harris, Jeffrey E. <u>Potential Risk of Lung Cancer from Diesel Engine Emissions</u>. Report to the Diesel Impacts Study Committee, National Research Council. Washington, D. C.: National Academy Press, 1981.

Hogan, William W. <u>Decision Analysis for Regulating Diesel Cars</u>. Report to the Diesel Impacts Study Committee, National Research Council. Washington, D.C.: National Academy Press, forthcoming.

McDonald, Roy and Gregory K. Ingram. <u>Diesel Car Regulation and Traffic Casualties</u>. Report to the Diesel Impacts Study Committee, National Research Council. Washington, D.C.: National Academy Press, forthcoming.